P9-CBV-595

Fun with Figures

J. A. H. HUNTER

Dover Publications, Inc.
New York

Published in Canada by General Publishing Company, Ltd., 30 Lesmill Road, Don Mills, Toronto, Ontario.

This Dover edition, first published in 1965, is an unabridged and corrected republication of the work first published by Oxford University Press in 1956. This edition is published by special arrangement with Oxford University Press, Toronto.

International Standard Book Number: 0-486-21364-1

Library of Congress Catalog Card Number: 65-17669

Manufactured in the United States of America

Dover Publications, Inc.
180 Varick Street
New York 14, N.Y.

Preface

Even the real mathematician may find something to amuse him in this little selection of teasers, but it is intended mostly for the regular run of folk who, like myself, take figuring rather less seriously.

For figures can be fun, as seems to be shown by the vast interest taken in such problems by thousands of kind readers, without whose ideas, support, and encouragement this little book might never have been published.

So now I take this opportunity of saying 'Many thanks' to all those good friends, known and unknown, and also to Eddie Phelan of *The Globe and Mail*.

J. A. H. HUNTER

Toronto
1956

Contents

1. Murder on the River   1
2. Cigars for Dad   2
3. How Old is 'Old'?   2
4. The Spider and the Fly   3
5. A Tale of the Subway   3
6. He Had a Good Memory   4
7. The Funny Figures One Sees   5
8. Trust a Postman to Know   5
9. How Many Cigarettes?   6
10. No Seven Here   7
11. Tim and His Brother Tom   7
12. Chicken Feed   8
13. No Catch in This   8
14. Kim's Camera   9
15. The Old Soldier   9
16. Ken's Pocket-money   10
17. Bicycles and Horses   10
18. Where Does the Money Go?   11
19. The Boss Wasn't Amused   12
20. How Many Marbles?   12
21. Three Girls at the Fountain   13
22. The Tale of Knok   13
23. North and South They Go   14
24. Those Licence Plates   15
25. So Little For So Many   15
26. The Car Game   16

27. Such a Waste 17
28. The Beauties of Kalota 17
29. The Hunters 18
30. Milly Miffen's Muffins 19
31. The Hot-rod Trials 19
32. How Green is the Garden 19
33. Time for Church 20
34. Betty's New Dress 21
35. Who's the Lady? 22
36. A Wise Wife 22
37. A Miracle at City Hall 23
38. A Wild Look in His Eyes 23
39. Payment Past Due 24
40. A Family Gathering 24
41. Jill Got Her Sundae 25
42. Roaches in the Warehouse 25
43. Buying Christmas Cards 26
44. Stamps at the Cigar Store 26
45. Simon Bought More Stamps 27
46. Five Threes and More 28
47. Mike's Morning Walk 28
48. Those Were the Days 29
49. Jill in the Pet-shop 29
50. Not a Square Deal 30
51. Cheaper for Two 31
52. The Road to Zirl 31
53. How Happy Would I Be With Either 32
54. He Gets It From School 32
55. It Was Hallowe'en 33
56. The Missing Miss 33
57. A Hopeless Case 34
58. Fair's Fair 35
59. Just a Simple Routine 35
60. A Man and His Sorrow 36
61. Three Men and Their Smokes 36
62. Horror in the Night 37
63. A Day on the River 37

64. A Man Came Back — 38
65. The Luncheon Party — 39
66. The New Apartment — 39
67. Change and Exchange — 40
68. So Soon Forgotten — 40
69. A Quiet Evening at Home — 41
70. Those Handy Cartons — 41
71. Tony Never Used To Be Like That — 42
72. It Fell From the Sky — 42
73. What a Tie! — 43
74. A Tale of Three Monkeys — 44
75. On the Queen's Highway — 45
76. A Game of Canasta — 45
77. Humpty Dumpty — 46
78. The Spreading Chestnut Tree — 46
79. Only Snails — 47
80. A Tale of the Cats — 47
81. A King Was Chosen — 48
82. Simon Bought Some Fruit — 49
83. Top Secret — 49
84. In Kalota There Was a King — 50
85. Mike's Pike — 50
86. One Rainy Day — 51
87. A Profitable Round — 52
88. A Monkey on the Job — 52
89. Three Friends and Their Birthdays — 53
90. It's Collars Today — 53
91. She Married Them Off — 54
92. The Beverage Room — 55
93. Digging for Dollars — 56
94. More Files for the Office — 56
95. Best Irish Linen — 57
96. Venice or Vladivostok? — 57
97. How Old is Jack? — 58
98. Kurt's Birthday — 58
99. The Story of Kalia — 58
100. Three Sisters — 59

101.	Spot the Winner	60
102.	They Played Checkers	60
103.	The Travel Set	61
104.	A One, a Four, and a Nine	61
105.	Shoes for Free	62
106.	That Was a Family	62
107.	John and His Bicycle	63
108.	Who Broke the Cups?	63
109.	Findings isn't Keepings	64
110.	Not for Children	64
111.	Which Road for Knokado?	65
112.	Sam Rides Home by Taxi	65
113.	For Betty's Birthday	66
114.	Only One Girl Among Them	67
115.	Their Birthday	67
116.	The Flatterer	68
117.	That Jack Again	68
118.	The Telephone Number	68
119.	Betty's Birthday	69
120.	Simple Simon Again	70
121.	A Parcel From Each	70
122.	That Old, Old Story	71
123.	A Glad Goodbye	71
124.	Some Christmas Shopping	72
125.	The Commuter	72
126.	One Fake Dollar	73
127.	Fun in the Tub	74
128.	Only Partly Right	74
129.	Neighbours	75
130.	The Ballad of Ballygan	75
131.	His Dinner Waited	76
132.	How Young is Young?	76
133.	The Art of Selling	77
134.	How High Was the Window?	77
135.	First Snow of the Winter	78
136.	The Expense Account	79
137.	Sal and the Skirts	79

138. *The Five Threes*   80
139. *A Trip to Neepawa*   80
140. *The Curse of Man*   81
141. *Peace at a Price*   82
142. *Doreen Got Her Puppy*   82
143. *So Many Bananas*   83
144. *On the Road*   83
145. *A Lady and Her Lingerie*   84
146. *Her Secret*   84
147. *The Marbles Rolled*   85
148. *Late, as Usual*   85
149. *But How Old Was Pam?*   86
150. *Cookies for Four*   86
    Typical Solutions
       A   89
       B   89
       C   90
       D   90
       E   91
       F   93
       G   93
       H   94
       I   95
       J   96
       K   96
       L   96
       M   97
       N   98
       O   100
  Answers   103

1
Murder on the River

'All aboard!' called Mary, stepping gingerly into the canoe from the slippery steps at the foot of the garden. She settled herself into the cushions while her husband started to paddle upstream; she yawned, she slept.

The noise of traffic overhead wakened her as they passed under the bridge some minutes later. She opened her eyes, looked over the side, and screamed: 'Steve, there's a hand in the water! It's a dead body!'

'You've been dreaming, darling,' he laughed, seeing what she saw. 'It's only an old glove full of nothing.' Relaxing with a sigh of relief, Mary closed her eyes again, and Steve went on paddling doggedly upstream.

After a further fifteen minutes he turned and made for home. 'Whoosh, whoosh, whoosh,' went his paddle— always the same steady rhythm as the light shell slipped through the water. They shot under the bridge, and then it didn't take long to make the remaining mile to their house.

As they arrived abreast their steps, Steve roused his wife with a shout: 'Wakey, wakey! Here we are, and there's your corpse.' And there indeed, still floating in mid-stream, was that very lifelike old glove which they had just overtaken.

Steve paddled at the same speed all the time. But what do you make the speed of the current?

2

Cigars for Dad

Simple Simon was in a spot: he wanted to give his Dad a dozen cigars for Christmas, and now he didn't know which to buy.

'They're so long,' he said, spreading out his fingers, 'with gold bands on them.' But the clerk only smiled. They all looked much the same, but Simon chose four of a cheap brand, four at double that price, and four at six cents more than the cheap cigars. But then he found that these would cost thirty-two cents more than he had in his pocket.

In the end he bought a dozen of the medium-priced cigars, instead of the four, and that just took all his money. So how much did he spend?

3

How Old is 'Old'?

Mr White loves kids, especially the noisy, rather grubby youngsters he meets going home in the street-car. The kids love him too, and always tell him the latest news from school.

Today they were full of their new teacher. When asked what she was like, they all agreed on one point: 'She's old!'

Amused by this, as he happened to know the girl who was an old school-friend of his youngest daughter, Mr White asked what they meant by 'old'. One said she must be twenty-seven, another suggested thirty-one, a third was sure she was thirty-six, and a fourth child went right up to thirty-nine.

In fact one guess was only one year out, one erred by three years, one was six years wrong, and one was as much as nine years from the truth. But could *you* give the teacher's age?

4

The Spider and the Fly

'Come right into my parlour,' said
   The spider to the fly,
'And answer one small question, please,
   Unless you want to die.
I've eaten scores of flies, of course,
   But tell me if you dare:
If females had two more, and males
   But half their present share,
How many flies like that, d'you think,
   I really would require,
To give me twenty-eight fly legs,
   The number I desire?'

5

A Tale of the Subway

He looked so forlorn, standing there all alone with the heedless, seething crowds milling around him. 'What's the big trouble, son?' I asked, my heart touched by his sad expression.

'I'm waiting for my brother Bob, but he's mighty late,' replied the boy. Cold shivers rippled down my spine as I pictured an even smaller lad lost somewhere in the vastness of the Toronto subway. 'That's bad,' I commented, 'but how old is he?'

'He's big,' was the proud answer, 'as big as you, Mister!'

When he spoke again it was in an even more confident tone: 'He's bigger than Mum, but not so old.' I smiled at that, and so must have encouraged the little fellow to continue. 'Dad said today that Mum's age in two years' time, divided by Bob's age now, and then added to Bob's age, comes to two-thirds of what Mum was when Bob was born.'

He paused a moment, a wicked grin spreading across his grubby little face. 'And today is Mum's birthday and also Bob's.'

So this was the little wretch on whom I had wasted my pity—this little demon with his tantalizing teaser that still keeps me awake at night! But maybe you can figure it out.

6

He Had a Good Memory

'Here's a dime for each year of your age, and a nickel for each year of Jill's,' said Uncle Frank, handing Jack some coins. 'Now go get yourselves some Easter candies.'

The boy started to thank him, and then remembered something. 'You gave us money the same way that Easter Dad was out west,' he told his uncle, 'but the other way round: nickels for my age and dimes for Jill's. I was as old as she is now.'

'So you've got a good memory,' chuckled Uncle Frank, 'but so have I, and that proposition cost me forty cents less.'

How much older is Jack than his sister?

7

The Funny Figures One Sees

Bert and Ben were strolling down Yonge Street when Ben stopped suddenly. 'Look at those funny figures,' he exclaimed, pointing across the road.

'Don't see anything funny about them,' rejoined his friend. 'They look all right to me!'

'Not those, you old wolf,' laughed Ben. 'I mean the number of that shop.'

Bert dragged his eyes away reluctantly and waited; he guessed what was coming. 'If you divide it by three,' continued Ben, 'there's one over; by five there's two over; by seven there's three, and by nine there's four over.'

'So what? There must be lots of numbers like that with this street running nearly up to Hudson Bay.' Bert was not in the mood for such nonsense.

Ben nodded. 'Yes, of course; but we're away down in the one-hundreds here.'

Bert wasn't co-operative, but maybe you can figure out that number.

8

Trust a Postman to Know

Myrtle Drive is a pleasant road of modern houses, seldom used by through traffic. Maybe its only fault is that all the houses are alike, but then that's merely a matter of opinion. Each doorway faces its twin squarely across the intervening hundred-and-twelve feet of lawns and road. By common agreement, the owners of these houses have put up no fencing between them to the front: a broad swathe of green grass runs right along each side between houses and sidewalk, broken only by neat flagged paths and driveways.

The postman comes to this road from the east. When

he is through, he delivers in Sylvan Avenue which runs across the western end of Myrtle Drive.

Today Mrs Smart, who lives at Number 7 and always watches for her mail, noticed the postman was following a different routine. When he reached her door she mentioned this. 'Yes, Ma'am,' he replied, 'I figured I was wasting shoe-leather in Myrtle Drive. So now I get around the job in the shortest possible distance.'

'But you went right down the other side today, and then back to the beginning of this side by a long diagonal, and now you'll have to finish this side,' observed the lady. 'Why don't you do it my way?' She explained her idea: to start at Number 1, cross the road to Number 2, then call at the neighbouring Number 4, then across to Number 3 followed by the neighbouring Number 5, and so on.

'But the distance would be exactly the same as by my own new routine,' the postman told her. And, bearing in mind that each house is built on a frontage of sixty-four feet, she had to agree with him.

So how many houses are there in Myrtle Drive?

9

How Many Cigarettes?

John gave up smoking last month—that is, he cut down considerably on his smoking. He explained that a complete break might be bad for his nerves, but his wife had heard that one before.

The first five days John kept well within the ration he had fixed for himself. The sixth day was not too good, however, for friends came visiting and he had to be sociable, but that day only brought his total for the six days up to his quota.

Next day was a Sunday and it rained continuously. Going to bed that evening, John was challenged by his wife: 'How's the quota now?' she asked. He had to admit he had smoked as many that day as on the previous day, seventeen cigarettes in fact, and that his daily average was now one over his ration.

What do you make his daily quota?

10

No Seven Here

This little teaser is really quite simple, but on second thoughts it isn't too easy at that. Five figures of the multiplication sum are shown, and the rest are indicated by crosses. One hint will be useful: no seven appears in the calculation. So now you may be able to find out what the final result is.

```
        2 x x
        3 x x
      -------
        5 x x
      x 4 x
    x x 3
    -----------
    x x x x x
```

11

Tim and His Brother Tom

Tom and Tim are brothers; their combined ages make up seventeen years. When Tom was as old as Tim was when Tim was twice as old as Tom was when Tom was fifteen years younger than Tim will be when Tim is twice his present age, Tom was two years younger than Tim was when Tim was three years older than Tom was when Tom was a third as old as Tim was when Tim was a year older than Tom was seven years ago.

So how old is Tim?

12

Chicken Feed

It was dusk. The last of the chores was done and Mike sat with his wife on the porch, looking out over their farm as the familiar scene melted into the deepening shadows. It was the time they always looked forward to: those peaceful moments together, resting after the day's hard toil.

But Mike was worried: the price of feed was up again. 'A sack of corn lasts our chickens only nine days,' he told Mary, 'and we won't be able to go on like that.' He puffed at his pipe: 'What say we get rid of them altogether?'

Mary's mind flashed to thoughts of her husband's favourite Sunday dinner. 'Surely you'll keep some for ourselves!' she exclaimed.

Mike considered the point a moment. 'Okay,' he conceded, 'I'll keep a dozen and sell the rest; and we'll cut the ration of corn by ten per cent and then I figure a sack will last thirty days.'

How many chickens did they have?

13

No Catch in This

Say! My digits are three,
But just what can they be?
Take a third of my first
    From a half of my third:
And a minus, right here,
    Would, of course, be absurd.

Now add twice my middle
To continue the riddle.
If you figured it right,

Then you can't have got seven;
For, believe me, I know
　　That you must have eleven.

There's no catch in this,
But don't take it amiss
When I add one thing more,
　　Just to make it quite clear:
You should know that a five
　　Is no part of me here.

14

Kim's Camera

Kim was showing Joe his new camera, explaining the advantages of coupled rangefinder, coated lens, flash synchronization, and other special features of his purchase. But he was particularly proud of having persuaded the dealer to let him have it at less than the regular price of $150.

'It was a funny thing,' he said, 'that I paid an exact number of dollars altogether.'

'Nothing very funny about that,' observed Joe.

Kim smiled. 'But I paid cash,' he told his friend, 'dollars, quarters, dimes, nickels, and pennies: the same number of each.'

That may have seemed funny to Kim, but it might amuse you to figure out what he paid for the camera.

15

The Old Soldier

Ron's grandfather is an old soldier, many times a veteran, for he enlisted way back in 1890. He was relating some of his adventures, and Ron sat there taking in every word. But after one particularly lurid episode the boy interrupted: 'Gee, that was exciting. But what year was it?'

The old man took a dollar from his pocket. 'You'll get this if you figure it out,' he replied. 'If you exchange the second and fourth figures you make a date ninety-nine years later than what I'm telling of.'

Ron couldn't do it in his head, but what would you say?

16

Ken's Pocket-money

First thing every Monday morning Ken receives his dollar; that's one thing the boy doesn't forget. By the beginning of last month he had managed to save a bit, but during the month his hoard dwindled away: each week he spent half of what he had had on the Monday after receiving his pocket-money.

When it came to the fourth Monday of the month, Ken had only $2.75 altogether when he counted his money at breakfast after receiving his weekly dollar. It was just too bad nobody had given him even a cent extra during the month; most times he could be sure of an occasional tip from one of his uncles.

How much did he have that first Monday of last month before his mother handed him that dollar?

17

Bicycles and Horses

They had been waiting a long time there in the car, but there was still no sign of Ronnie's mother. Uncle Frank turned to his small nephew: 'We'll play a game,' he said. 'I'll give you a nickel for every bicycle that passes, and a dime for every car; and you give me seven cents for every horse.'

Traffic was light that day, but it promised to be a profitable deal for the boy. When his mother emerged from the store, bringing the game to an end, they had counted only half as many horses as cars. Ronnie didn't want to stop: he had won sixty-seven cents already.

How many bicycles passed by?

18

Where Does the Money Go?

Mrs Brown goes shopping most days, and most days she returns home in a complete muddle as to what she's spent. Yesterday, in the first store she spent half of what she had in her bag. In the next store she spent half of what remained, and so she then called in at her bank and cashed a cheque for ten dollars.

Out in the street again, however, she found it had started to rain. So she slipped into the drug-store across the way and made one final purchase before going home: for that she paid the same amount as she'd spent in the first store.

'Now for a taxi,' she thought as she emerged from the drug-store, for there was a veritable deluge outside. But that meant checking her cash to be sure she had enough, and so Mrs Brown looked in her bag and found just five dollars and forty cents there. 'I'm sure I don't know where it all goes,' she grumbled, but she hailed a passing taxi nevertheless.

Of course it's easy to be wise after the event, but how much would you say she'd spent?

19

The Boss Wasn't Amused

Jim's a crank, one of those peculiar people who enjoy teasers and crossword puzzles. And the worst of it is that he always wants others to join in the fun.

Today when his new boss asked him his age, Jim answered: 'If you had two or three times as many dollars as I would have dimes if I had twice as many nickels as my last boss would have if he had three times as many cents as my age in years, and if none of us had any coins at all, you would have three dollars more than twice as many as I was years old five years ago.'

That's what he said, and tomorrow he'll be looking for a new job. But how old is Jim?

20

How Many Marbles?

'Break it up!' called their father, coming out into the yard where the three boys were playing marbles. 'Dinner's nearly ready!' But he relented when he saw the looks on their faces. 'Okay. Two minutes more, but that's all.'

They continued their game with fierce concentration under his critical gaze. All too soon the time was up. 'Come along now, no more.' And now he meant it.

'When you came out I'd lost as many as Tim had won,' grumbled one of the boys, 'and now we stop just as my luck is turning.'

'But Tim started with half of what I had, and Tam started with five times as many as me,' piped Tom shrilly, 'and now Tam has as many as I had when you first came out.'

The third boy had pocketed his marbles. 'I've got three less than Tam had when we started,' he told them.

'Anyway, two of you ended up equal,' commented their father, leading them into the house.

How many marbles did the boys have among them?

21

Three Girls at the Fountain

The girls sat at the fountain, sipping coffee while they discussed that memorable party where they first met. It was one of those 'D'you remember?' conversations, and they only disagreed about the date.

Mary said it had been Thursday, May the 8th; Gwen agreed on the month, but was sure it was a Tuesday and on the 10th; Jane insisted it was June the 8th, and Friday at that.

The argument became quite heated and even involved the soda clerk, who told them his birthday was on April the 1st and that it had been on a Tuesday that year.

In fact none of the girls was completely right, but among them they had mentioned the correct month and the correct day of the month and the correct day of the week. One had made only one true statement, one had been wrong on only one point, and one had been completely wrong.

Assuming the soda clerk was right, what was the date of that party?

22

The Tale of Knok

'There was a man in the land of Kalota whose name was Knok; and he was mighty of stature and a cunning worker withal in brass and copper. And Knok had in his place many stones for the true weighing of his wares: stones of one kalen and other stones of greater weights.

'And it came to pass that in the season of the rains great floods spread across the land, and the people fled to the high places leaving all behind them.

'And after a space, when the waters had receded, the people returned; and there was much lamentation, for the homes of the people were destroyed.

'And in all the rubbish Knok could find only one of his stones, a great stone which weighed forty kalens. And Knok took a mighty hammer and he smote the stone and clave it into four pieces, and no piece was like unto another piece. And with the four pieces was he able to weigh his wares, and the people paid him for each kalen of weight, yea for each kalen from one up to forty kalens.

'So Knok wrought for seven days and seven nights, making for each the wares which he desired, and he waxed rich in the land.'

That is what the Book says. But what did those four pieces of stone weigh?

23

North and South They Go

'Hello, Ken. D'you think you're a traffic cop?' called Uncle Frank, coming up to where the youngster stood making notes on a piece of paper as the cars streamed by.

Ken checked carefully before answering. 'Seven hundred and eighty cars have passed while I've been here,' he informed his uncle, 'and if a quarter of the cars going north had been going south, and if a fifth of the cars going south had been going north, and if a sixth of the cars which would then have been going north had not been on the highway at all, there would have been ten more cars going south than north.'

That hardly sounded like a police officer, but anyway

Uncle Frank did manage to figure out how many cars had been checked going north. What do you make of it?

24

Those Licence Plates

Paul looked down at the shiny new licence plates on the table. The three friends had gone together to renew their car licences and were now having some refreshment. 'It's all nonsense having new plates every year,' he remarked. 'Just as I begin to remember my number it has to be changed.'

'That goes for me, too,' agreed Dick, 'but there's something funny I've noticed about our numbers this time. Your first figure is the same as Hal's last, and your last the same as his first, and you've both got the same figures in the middle.'

'And there's something else,' exclaimed Hal. 'Your two numbers add up to mine.'

Paul picked up the check, which his friends seemed too engrossed to have noticed. 'Let's go,' he said, 'but I would have you note that Dick's two middle figures are the same and are also the same as your first figure, and his first figure is the same as mine.'

Well, it certainly was a mix-up of figures among them; fortunately all three had been given 4-figure numbers. So now what was Dick's licence number?

25

So Little For So Many

'Lordy, but I'm tired,' sighed Sue as she and Sal settled down in the street-car. 'The same old skirts every day.

About fifteen dozen for the whole week, and I hate the sight of the blessed things.'

'What about me?' said Sal. 'Yours are quick to make and you get twenty-seven cents for each.'

The girls were on their way home that Friday evening from the garment factory where they were employed as operators on piece-work. As the car jerked its way from stop to stop, they compared notes on the week's work. It transpired that there was a difference of only a nickel between their earnings that week, although Sal was paid forty-seven cents for each of the dresses she had made.

What do you make Sal's earnings that week?

26

The Car Game

Jack and Jill sat at the living-room window playing the car game. She paid him a nickel for every car that passed from right to left, and he paid her the same for every car passing the other way.

They live in a quiet road with little traffic, but most of the cars seemed to be passing in Jack's favour. 'It isn't fair,' complained Jill at last. 'Only seven have gone my way.'

'Okay, sister,' replied Jack, who was becoming rather bored with the game. 'We'll stop after five more have passed.' But of those next five cars only one favoured him, and so he ended up just one nickel to the good.

How many cars passed during the game?

27

Such a Waste

'What happened, dear?' asked John, as his wife came into the living-room laden with parcels and obviously rather upset.

'Everything went wrong! The store was like a mad-house, and then I had to stand all the way back in the street-car and a bag burst and I lost all my oranges.'

Gwen sounded very near to tears, and John tried to console her: 'What's a few oranges, anyway?'

'But it's such a waste,' she sighed, 'and they were a nickel each.'

It would have been more serious if she had lost the bananas, of which she had bought more and paid as many cents each as the number she bought. As Gwen paid just eighty-nine cents altogether for the oranges and bananas, you can probably figure out how many oranges she bought.

28

The Beauties of Kalota

John seems to have had a thoroughly good time last year when he was visiting in Kalota. He did find it difficult, however, to understand the Kalotan women. It is the strange custom of that island that a woman must never make two consecutive true or untrue statements; if one statement is true, then her next is a lie, and vice versa.

The hospitable merchant with whom John was staying had four attractive daughters: Kassa, Kessa, Kissa, and Kossa. John says he tried to find out their ages, but the girls got him really tangled, even though there are no twins in Kalota.

Kassa started it: 'Kissa is twenty-three, and Kessa

twenty-two.' Kessa's version was quite different: 'Kossa is twenty,' she told him, 'and Kissa is twenty-two.'

John appealed to Kissa, for whom he appears to have fallen in a big way, but she wasn't very helpful. 'Kassa is twenty-two, and Kossa is nineteen,' she assured him demurely.

John knew very well that one of the girls was nineteen, and so he should have been able to figure it all out from what they had said; but then he never was very smart. What do you make their ages?

29

The Hunters

Towards the end of a day's hunting, Jim and Joe were resting on an old log, enjoying a smoke. Comparing results they found each had expended the same amount of ammunition: Joe, by no means a good shot, had averaged three shots for every bird he had brought down; Jim, who admits he only goes hunting for the sake of the evenings in camp, had only killed one bird for every four shots.

Suddenly, as they talked together, a flight of duck appeared, coming in low in their direction. Each fired once, and after that they saw no more birds that day.

Trudging along some time later, tired and very thirsty, Jim remarked on the fact that he had killed the same number of birds as his friend. 'So what?' chuckled Joe. 'It's also funny we neither of us killed two birds with one shot today.'

How many did Jim kill?

30

Milly Miffen's Muffins

Milly Miffen made a muffin more than Molly's mother made and Milly Miffen's mother made a muffin more than Molly made; and Milly, Molly's mother, Molly, and Milly's mother made fifty muffins, but Milly and Molly's mother made four muffins more than Molly and Milly's mother made. So murmur now how many muffins Milly made.

31

The Hot-rod Trials

Larry looked up from the sports page as his father entered the room. 'I don't get this, Dad,' he said. His father looked at the paragraph which described some hot-rod trials down on the salt-flats; speeds were checked over a six-mile run, it said, the cars being timed at 3 miles, 4½ miles, and 6 miles. 'What's wrong with that?' he asked.

'Well,' Larry told him, 'it says one car averaged 140 for the first 3 miles, 168 for the next 1½ miles, and 210 for the final bit. I make that an average of just under 173, but that's not what the paper says.'

What do you make the average speed for that six-mile run?

32

How Green is the Garden

Spring was in the air, and Gus was in the garden marking out the new lawn.

'I don't like the shape: it's too square,' commented his wife. 'But if you make it six feet longer and three feet narrower, there'll be room for a pathway up the side. Besides, as the area will be the same, you won't need extra seed.'

But her mother had other ideas. 'Who wants a path?' she asked in that tone which made Gus squirm. 'Make it eight feet shorter and six feet wider and you'll still get the same area of lawn and there'll be space for vegetables at the far end.'

There was no arguing against that, so now Gus spends his evenings hunting slugs and caterpillars. But what was the area of the lawn?

33

Time for Church

'Ding!' pealed the bell of St Mary's, its silvery tones ringing high and clear. 'Dong!' boomed the deep throaty voice of St Mark's in reply, but just seven seconds later. And then they were silent.

Kit, relaxing lazily after a hearty Sunday breakfast, reached for a sheet of paper and began to scribble; ever since the bells had started simultaneously less than an hour previously, he had been amusing himself keeping count of their peals. They had tolled at regular but different intervals: one exactly a hundred and seven times and the other sixty-eight times.

'Come along, dear,' called Kate from the hall, 'we mustn't be late again.' 'Just a moment while I finish this,' replied her husband, not deigning to look up. But Kate wasn't standing for any such nonsense. 'You and your problems! You can finish it after church,' she told him, rushing into the room and grabbing the paper from his hand.

So Kit donned his hat and accompanied his wife dutifully. He hadn't actually timed those bells, but somebody had told him that each pealed at intervals of an exact number of seconds. He couldn't help thinking about that timing, especially during the sermon; but it wasn't until he was back home that he figured out the respective intervals of the bells.

What do you make them?

34

Betty's New Dress

It was late Saturday afternoon. Mike jumped up from his chair as his wife came in. 'Let's see the new dress, dear,' he said, helping her out of her coat. 'But I didn't get it,' she told him, 'and now it's too late as they'll have closed.'

'You went specially for it! What happened?' Mike was puzzled, but Betty explained. 'I stopped in at Jenkins on the way and got some nylons,' she said, 'and then I hadn't enough for the dress.'

'But they know you well enough. Surely you could have paid the balance on Monday?'

Betty shook her head. 'No, they were plain mean. I offered to pay nine-tenths down, just what I had in my bag, and the rest next week. But they wouldn't do it. The girl said the price would be upped fifteen per cent on Monday and that they'd have no difficulty in selling it.'

There aren't many husbands like Mike. First thing Monday morning he was at the shop and bought the dress. The price had gone up, as expected, and he had to pay $7.15 more than the down-payment his wife had offered two days before, but that didn't worry him.

What did Mike pay?

35

Who's the Lady?

'Saw you with a couple of good-lookers yesterday,' remarked Mike with a sly grin.

'What, Gwen and Peg? Why, one of them's my niece.' Steve hesitated a moment, chuckling. Then: 'When my niece is as old as I am now, Gwen will be twice as old as Peg is now. And I'll gladly introduce you to my niece if you'll tell me her name, even though they're both in their teens.'

'You old wolf!' laughed Mike, who knew very well that his friend was just forty. But he got the introduction.

What do you make of it?

36

A Wise Wife

'Listen, my dear,' remarked Mr Jones one evening as they sat in the living-room after supper. 'This is really interesting.'

Mrs Jones could never understand her husband's morbid interest in figures, but after four years of marriage she knew it was wise to humour him over his strange hobby. She looked up from her reading with a smile. 'Tell me, darling,' she begged.

'Well,' he told her, glancing at a piece of paper in his hand, 'I've added together the year you were born, the year we were married, and your present age. It makes a total of thirty-nine hundred, which is exactly a hundred times my age.'

Mrs Jones seemed to be suitably impressed. 'What an amazing thing!' she exclaimed.

In what year was this scene of domestic felicity enacted?

37

A Miracle at City Hall

Vic stopped suddenly, much to the audible annoyance of the crusty old gentleman (apparently a retired colonel, from his flow of language) who was walking close behind. But Vic was far too overcome with emotion to pay much attention to mere man in the face of this miracle.

He looked at his watch, and then again at the great clock of City Hall. It was indeed true; he had not been mistaken. The clock was actually showing the correct time!

Vic hastened home to tell his good wife the astonishing news. A little later, telling the tale as calmly as he could, he remembered that the minute hand had been exactly three minute-divisions ahead of the hour hand, and that both hands had been precisely at minute-divisions.

At what time did Vic witness this modern miracle?

38

A Wild Look in His Eyes

'For heaven's sake! What's come over you today?' exclaimed Mrs Green, dropping the carton of cream into her bag. 'I only asked the price.'

'Yes, Ma'am, and I'll tell you again,' the clerk replied, a wild look in his eyes. 'It's a quarter of the price of a quart, and a quart and a quarter would cost you three quarters more than a quarter of the price of a quarter and a quart.'

At that moment a couple of uniformed men rushed into the store and hustled the poor clerk out to a waiting ambulance. But what was the price of that half-pint of cream?

39

Payment Past Due

'Just a moment! You wait!'
Shouted Alf, quite irate,
    At Doug, who'd come dunning for dough.

'If you'd invoiced us right,
We'd have paid you on sight:
    It's all your own fault we've been slow.

'You'll remember we bought
More than ten of one sort,
    At eighty-nine cents: that you know.

'As regards all the rest,
Which you said were your best
    (And priced, you'll admit, not too low),

'We bought fifteen or more
At a dollar and four,
    For that was as high as we'd go.

'So the total should be
Thirty-one thirty-three.
    I'll pay you right now what we owe.'

How many of each sort had they bought?

40

A Family Gathering

It was a family reunion dinner—just the six of them:
Betty with her sister Mrs Armstrong, and her brother,
and of course their respective spouses.

Mr Armstrong sat beside Amy, while Alf faced Mrs

Cross. Charles was placed on the left of Betty's sister, while Betty sat between Bert and Mr Briggs.

Clara started them all laughing when she said: 'There'll be no fighting, as wives and husbands aren't sitting together.'

Now you can figure out the first names of those three couples.

41

Jill Got Her Sundae

'How much have you got?' asked Jill as she went with her brother into the drug-store.

Jack grinned. 'I've only got nickels, dimes, and quarters: but fifteen altogether,' he replied. 'Yes, that's all I've got,' he went on after checking his pockets again, 'and if the dimes were nickels there would be four times as many nickels as quarters, but if the nickels were quarters I'd have twice as many quarters as dimes.'

He jingled the coins, laughing at the expression on Jill's face. 'And now I'll buy you a special sundae if you guess how much I've got.' And how much was it?

42

Roaches in the Warehouse

Scuttling around the warehouse on Sunday morning, Ron met his old friend Roy hurrying towards him. 'No time for gossip now,' called Roy, waggling a shiny wing-case as he passed. 'We'll meet the other side.'

The two roaches continued on their respective ways, close up to the foot of the wall. But Ron turned back some moments later. 'What the heck?' he thought. 'I'll get

back home.' After retracing his steps a matter of only
ten feet, however, he changed his mind about this: Roy
would be so worried if he didn't turn up on the other
side as expected. So Ron turned once more and continued
his progress in the original direction.

Maybe you don't know that old warehouse down on
Front Street? It is just forty-four yards around the floor,
and Ron is the champion sprinter in his pestilential little
community. Roy isn't slow, but at their normal regular
speeds he does only eight feet to Ron's eleven. On this
occasion they both scuttled along at their normal regular
speeds from the moment when they first passed each other
to the time when they met again on the other side of the
place, and of course Ron didn't waste any time on either
of his two 'turns'.

How far had Roy travelled when he did meet Ron at
last on the other side of the warehouse?

43

Buying Christmas Cards

Having been dragged into the crowded store by their wives
to help choose Christmas cards, Les and Len were very
glad to complete their purchases and escape to a nearby
restaurant for refreshment.

As they all sat sipping their coffee, they started comparing their cards. Quite by chance it turned out that
each couple had bought twenty: Les had bought three
more than Ann, and Elsie had bought only four herself.

It may seem quite irrelevant, but which of the two
girls was married to Len?

44

Stamps at the Cigar Store

'Did you remember the stamps?' asked Sally as her son
came into the living-room. 'Sure, Mum,' replied Jack,

'but I didn't go anywhere near the post office. The girl at the cigar store let me have some, but she didn't have any of the 7¢. I got four different values, some of each.'

'So what values did you get, then?' his mother inquired.

'That's for you to figure out, mother darling,' the cheeky little wretch told her, 'as one of each value would make exactly a dime. There are twice as many of the highest value as of the lowest, and you owe me 31¢ for the eleven I got.'

'You no tell, I no pay,' laughed Sally, making no attempt to humour him. From what Ken had said, she could have figured out how many 2¢ stamps he must have bought. How many would you say?

45

Simon Bought More Stamps

When Simple Simon worked in that office down on Peter Street, supposedly making himself generally useful doing odd jobs, he was often sent out to buy stamps.

Three days before he lost the job he had to get stamps of two denominations, ten of one and six of the other. For these he entered an expenditure of $1.00 in his little petty-cash record. The following day he went to the post office again for stamps of the same two values, but he had to buy five of each and he charged the firm 45¢ as his outlay. His last day there he was sent off again, this time for six of one denomination and twelve of the other, the same values as he had got the day before; he entered his expenditure as $1.14.

Unfortunately for Simon, his boss was in a bad mood when he examined the stamp account in which the boy had entered the amounts. 'One of these entries must be wrong,' he said, pointing to the record of the last three purchases. And so poor Simon lost his job.

Can you spot the wrong entry?

46

Five Threes and More

Here we have a regular but not too simple division-sum. Five of the threes are shown, and the other figures are represented by crosses; there is at least one '5' in the calculation. It all looks far more difficult than it really is, so now you are asked to find the 5-figure number.

```
x x 3 ) x x x x x ( x x x
        3 x x
        ─────
          x x x
          x 3 x
          ─────
            x 3 x
            x x 3
            ─────
            - - -
```

47

Mike's Morning Walk

For the past few weeks Mike has been going to the office by street-car; it stops just by the end of his road and is really quite convenient. His wife is delighted to have the use of the automobile, and thinks him the most generous of husbands. Only to himself will Mike admit that parking troubles were the main reason for his new routine.

Walking along the sidewalk with his regular strides, Mike noticed today that he takes two paces to each of the big concrete slabs. When he lengthens his pace by six inches he takes five strides to every three slabs.

So now you can figure out the length of his regular stride.

48

Those Were the Days

Gil Giles is the biggest farmer around here, but he is never tired of telling how he started. 'I was a hired hand over with Sam Miller's dad until I married Mrs Giles, he'll say, chewing the end of a 50-cent cigar, 'and then I rented a little place and used my savings to get some stock.'

Mostly he pauses at that point to make sure he has your attention. And then he'll go on: 'Things were very different in those days. I bought pigs at nine dollars, geese at seventy cents, and chickens at fifty cents: a hundred head altogether and they cost me just a hundred dollars.'

Well, he's certainly worked hard to get where he is now. But can you figure out how many geese he bought?

49

Jill in the Pet-shop

It was a very small and smelly pet-shop, but Jill was quite fascinated. Having moved into a new apartment where animals were banned and children only just tolerated, she was thrilled to have found this little haven so close to her new home.

The birds, somewhat bedraggled in their cages, didn't interest her so much; but she loved the puppies and the kittens. The proprietor had been watching her hopefully. 'Perhaps your dad will buy you one,' he ventured at last.

'No, it's not allowed at home,' replied Jill. 'But may I come in to see them sometimes, please?' She smiled, adding: 'I can bring food for them if you'll tell me how many you have.'

The old man loved children almost as much as animals. 'Come in as often as you like, young lady,' he told her, 'but don't trouble to bring food. I have eleven more animals than birds. But if I had as many animals as I have birds, and as many birds as I have animals, there'd be only four-fifths as many legs altogether as there are now.'

Jill thought that sounded sort of funny, but it stuck in her mind. Later that day she told Jack about the shop and repeated the old man's words, and Jack figured it out for her.

How many birds were there in the pet-shop?

50

Not a Square Deal

Mr Kemitz groaned when the old man walked into his office: he'd have to work for his money this time. For this was Mr Kirby, eccentric but very rich. And he came straight to the point, telling the realtor he wanted to buy a rectangular lot which must comply with certain explicit conditions.

'Four times the frontage, added to three times the depth, must come to exactly a hundred and twenty yards,' he said, 'and each must be an exact number of yards at that and I want the greatest possible area.' And with that he stalked out of the office, leaving Mr Kemitz with no mean problem.

But the lot was found and the price was high and Mr Kemitz wouldn't mind a few more like that. But what do you make the area of that lot?

51

Cheaper for Two

'Sorry I've been so long, dear,' remarked Ruth as she came out of the hardware store and got into the car. Her husband was a patient man, but sometimes unexpectedly observant. 'That's all right,' he told her, 'but I thought you only wanted a kettle.' He put the parcels on the back seat. 'Why buy a skillet too when you have several at home?'

'Oh, but it was such a bargain,' explained Ruth. 'If I'd bought them separately, five-ninths of the total cost would have been on the kettle, and I would have had to pay a dollar thirty more altogether.' She flashed her husband one of her most winning smiles. 'So you see they really let me have the skillet for two dollars ten.'

You can't defeat that sort of argument, and anyway a wise man's wife is always right. But what would the kettle have cost by itself?

52

The Road to Zirl

'Now what do we do?' asked Jim, stopping the car where the road ended at a deserted quarry. 'You're the navigator!'

His wife looked up from the road-map. 'We must have gone wrong at that fork an hour ago,' she told him. 'You remember it said "Zirl 12 miles", but we weren't sure which way it pointed.'

Jim leaned back and lit a cigarette. 'I guess you're right at that,' he laughed, 'but we've been driving four hours and I figure that's eighteen miles more than we should have done.' He started to turn the car. 'Anyway, it's your responsibility.'

They reached Zirl without any further difficulties, arriving not too late for an excellent little dinner in the quaint old hotel. Assuming that Jim drove at a steady speed the whole time, how far is Zirl from where they had started out that day?

53

How Happy Would I Be With Either

'Come along, Garry, make up your mind!' cried his mother impatiently as the boy stood there eyeing the candies. Garry sighed. It was always like that when he took his mother shopping.

He only had a few coins to spend, and it was really very difficult to make a quick decision on such an important matter. 'I've got just enough for four ounces of those nut creams,' he said, 'but the chocolate fudge is cheaper and I could just buy twelve ounces of it.'

His mother was coldly practical. 'Get both,' she told him, 'the same quantity of each, and don't waste any more time.' And that is what he did, spending all his money on the purchase.

How many ounces did he buy altogether?

54

He Gets It From School

They've started a craze for teasers at school, and now it's almost impossible to get a straight answer out of Larry on any question at all. Today his mother asked the boy how much he had spent for her at the hardware store, and his reply was a good example of the way he is these days.

'If the check had shown half as many dollars and twice as many cents,' said Larry, in seeming seriousness, 'I would have spent two dollars fifty-eight less than I'd have spent if it had shown twice as many dollars and half as many cents.'

Now what could the poor woman make of that?

55

It Was Hallowe'en

'There they start. Now we'll have no peace,' growled Len, looking across the room to where his wife was fiddling with the radio. Ann smiled. It was the same every Hallowe'en, the same grumbles, but she knew very well he'd let no child ring the bell in vain.

'I'll go,' she said, picking up her bag and going out into the hall. She returned a moment later, however. 'Give me some change, darling,' she asked him. 'If I gave them each a dime I'd have a nickel over. So give me a quarter and then I can just give them each fifteen cents.'

Len didn't ask how many kids there were in the group outside the door, for he'd figured it out almost before his wife had finished speaking. But what do you say?

56

The Missing Miss

They were arguing about those two very pretty girls who had been at the party last Christmas. John had referred to them as Mary and Jane, and that started the argument. 'But that's wrong,' said Jim. 'Their names were Elsie and Gwen.'

Joe agreed with Jim as regards Elsie, but he thought the other girl was Ann. Jack was in no doubt about the names. 'I saw far more of them than you did,' he said, 'and of course they were Ann and Mary.'

Among them the four boys certainly did mention the correct names. Jack gave one name right but the other wrong, and so did two of the other three. The fourth boy was completely wrong.

The strange thing about all this, however, was that there had been no girl at all at the party with one of the names mentioned. But what name would you say that was?

57

A Hopeless Case

Professor Brayne is so absent-minded. They say he was seen one wintry morning strolling down the street with thick wool mittens on his feet, carrying his shoes.

In the parking lot on one occasion he couldn't locate his car; he'd forgotten the number and even the make. 'It's four figures,' he told the man, 'and the first is two more than the last. And they're all different and there isn't a zero.'

The Professor paused, but then remembered something else. 'You'll find the second figure is more than the third; and it's all the exact reverse of my own telephone number.'

'And what's that, sir?' asked the bewildered attendant.

'Unfortunately I've forgotten that too,' replied the old gentleman, 'but I remember noticing that the two numbers add up to 16,456.'

'If you can remember all that—' but he didn't finish the sentence: the attendant didn't want to hurt anybody's feelings. But he figured it out correctly, and so can you.

58

Fair's Fair

Uncle Frank had sent Jack a lot of foreign stamps. Now the boy had them all out on the rug and was sorting the stamps with his sister's help. Two of them were badly torn, so they were thrown away, and then Jill claimed her share.

'He didn't say how many were for you,' Jack told her, 'but I'll give you a number which, multiplied by itself, makes half of all he sent us.'

'That's big of you,' cried Jill, grabbing a good handful of stamps and making for the door, 'but it would still leave you just eight-ninths of what he sent, you greedy beast!'

So how many stamps did Uncle Frank send?

59

Just a Simple Routine

If you follow my rule with fifteen,
Then you'll surely get seventy-four.
The routine can then also be seen
With what number you wish—say a score.

With a score you get seventy-one,
But with five just an eighty you'll see.
So perhaps you can figure for fun
What the answer with thirty will be.

60

A Man and His Sorrow

'You pay me, Madam,' remarked the clerk, handing her
the parcel done up in crisp green paper. Mrs Ryle gave
the man a 20-dollar bill and glanced at the check in her
hand. But something was wrong. 'I think you've made
a mistake,' she said. 'You've charged me a dollar sixty-
two too much.'

The clerk sighed. It was almost impossible to con-
centrate when every thought was of his Tina, of the
happiness which had been theirs and of the miserable,
lonely years ahead. 'I'm very sorry,' he mumbled humbly,
looking at the check. 'I must have written down the cents
as dollars and the dollars as cents.'

Mrs Ryle could see the sadness in his eyes, but what
could she do? 'You also reversed the dollars when you
wrote them down as cents,' she told him, 'but we'll get
it right now.' She spoke kindly, for she also had known
unhappiness.

How much was the correct change out of that bill?

61

Three Men and Their Smokes

Jack accompanied his friends when they went into the
cigar store. Seeing his favourite brand on the shelf, he
bought some cigars at 25¢ each. Jim bought a cheaper
brand at 20¢, and Joe chose cheroots at only 15¢.

One of the three bought nine 'smokes', while the other
two bought five each. Among them they spent an exact
number of dollars.

Can you figure out which of them bought nine?

62

Horror in the Night

I was crouched in one corner of that vast bare hall, alone with my fear—but not alone, for all about me were the voices, whispers full of menace, racking my very soul with their strangely familiar questions. 'What did Mary pay?' they asked, and 'How spake they in Kalota?' and 'When did Ken return home?'. There was no respite. The air was filled with questions. I must answer, but I knew not the answers to those chorused and peremptory demands. It was horrible.

In that great hall, its floor and ceiling a dull, dead black, its longer walls a deep blood-red, its shorter end-walls a sickly yellow, my mind knew only one number: ninety feet; the length of the chamber was branded in my fear-crazed brain. The width and height of the place were less, but that was all I knew.

Then the voices were no longer there. Looking up, close into the most distant corner of that hall, I saw a Thing: a monstrous many-legged horror bathed in pulsing green light. 'You have just one last chance, you puerile problemist,' it thundered. 'If I crawled to you now, creeping ever nearer by the shortest possible route to tear you limb from limb, which colours would I traverse?'

I froze, speechless with terror. But the Thing continued: 'Figure it out if you would live.' Those writhing legs began to show a purpose in their weavings. I screamed, and found myself awake, still screaming. But what was the answer?

63

A Day on the River

It's a pleasure to take a trip on the river in the little ferry which runs up and down between Tormill and Torbridge.

Stretched out in a deck chair, watching the low green banks slip by, Hank felt at peace with the world. He had bought a return ticket and was now on the way back home to Torbridge.

The run up-river had taken just one-and-a-half hours, and he had been told the return trip would take forty minutes 'if all went well'. Lazing there half asleep, he found the steady beat of the engine most soothing. 'Chug, chug, chug,' it never varied; the same smooth rhythm all the way up to Tormill that morning and now still the same.

But suddenly there was silence, a hush which brought Hank out of his chair. Calling up to the tiny wheel-house in some alarm, he asked what was amiss. 'Engine's dead,' was the laconic reply, 'but we've only three miles to go and plenty of water. We'll drift it in thirty-six minutes.' And so it was.

How far is it from Torbridge to Tormill by river?

64
A Man Came Back

It was early in May 1952, and Beryl and Bob were spending a quiet Sunday afternoon sitting in the garden, enjoying the peaceful scene. Close at hand the children were playing: Garry, and Doreen who was exactly three-quarters her brother's age.

'It's strange to think back to this day just seven years ago,' mused Bob. 'I was over there in Bremerhaven, probably sitting on a heap of rubble, far, far from home.'

Beryl took his hand and squeezed it. 'And I was here,' she whispered, 'counting the days to your return. And even Doreen, only half Garry's age, seemed to sense my relief and happiness.'

In what year was Doreen born?

65
The Luncheon Party

'We've all remarked on it many times before,' observed Mr Bankes, 'but it still gets me. Where else in the world would you find a lawyer, an author, a dentist, and a banker at one table, and bearing the names of Law, Penn, Bankes, and Tooth?'

'It's worth sending in to Ripley,' replied the dentist, whose surname corresponded to Mr Penn's profession, 'especially as our names don't agree with our respective occupations.'

It was indeed rather strange. But what do you make Mr Tooth's occupation?

66
The New Apartment

'We don't want to live so far out,' objected Helen when her husband showed her the advertisement of that apartment in St Catherine Street. 'But it sounds just what we're looking for,' replied John, 'and the rent's quite reasonable.'

They had just arrived from Halifax and had checked into a downtown hotel with the idea of moving as soon as they could find what they wanted. And Helen did want to be fairly central. 'Just look at the number— up in the four-thousands,' she insisted, pointing to the paper. 'That must be miles out.'

But John laughed. 'You mustn't be misled by that,' he said. 'They number the houses here by the block system.' While he was explaining this, his wife doodled aimlessly on the margin of the paper, quite prepared to accept his word without a lot of talk.

Suddenly she interrupted him. 'That's funny,' she cried, indicating her scribbling. 'If you shift the last figure

of the number of that house and put it in front of the
first figure, and then add the new number to the original,
you get seven nine three one: that's your old home number
followed by mine.'

This was an omen not to be disregarded. They went
right to the place, and liked it so much that John rented
it there and then. But what do you make the number
of the house?

67

Change and Exchange

Mike pushed a handful of coins through the grill and
asked the teller to let him have bills in exchange. 'How
much is it?' asked the girl, but Mike wasn't sure. So
now she had to count the coins: some customers want
everything done for them.

Several disdainful sniffs later, the teller slipped fifteen
cents back through the grill to him. 'There were as many
pennies as nickels and dimes together, but twice as many
dimes as nickels,' she told Mike, 'and now three-quarters
of the value is in quarters and half-dollars.'

Sweeping the coins into a drawer, she handed him one
crisp new bill. And Mike had the grace to thank her
for her trouble. But what was the value of that bill?

68

So Soon Forgotten

The girls were discussing three boys they had met on
their summer vacation. There was no doubt at all as to
the first names, of course, but they just could not agree
about the surnames of the young men.

Peg said she had spent a lot of time with Jack Lennox, but she had not liked John Everard. 'I quite agree with you about John Everard,' Pam nodded, 'but my impression is that Jack's surname was Best.'

Now Pat had her say, insisting that the Best boy was John, and the other two Jim Lennox and Jack Everard.

As it happens, each of the girls was right about at least one boy, and the surnames were certainly the three they mentioned. Can you link the names correctly?

69

A Quiet Evening at Home

Len and Les spend many a quiet evening playing checkers. They play for cigarettes, the loser handing over one at the end of each game.

Last Tuesday they had a long session which resulted in one of the two friends winning a net total of ten cigarettes from the other. At the finish they each had six cigarettes, although Les had smoked three and Len seven during the evening's play.

How many cigarettes do you figure Les had when they started playing that evening?

70

Those Handy Cartons

Tam, Tim, and Tom were sent out to buy the weekend supply of milk. They were told to get seven and a half quarts. On their return, the boys trooped into the kitchen and stacked an imposing collection of pint and quart cartons on the table.

'But why so many pints?' asked their mother in some

surprise. 'You know it costs more that way.'

'You didn't tell us,' replied Tam, 'so we each brought back the same quantity of milk.'

'But we each got a different number of cartons,' said Tim. 'And I bought the most pints,' added Tom.

Can you figure out how many pint cartons Tom did buy?

71

Tony Never Used To Be Like That

Tony used to be such a polite little boy, but now his mother says he must be picking up bad habits at school. A few days ago, for instance, he was really rather cheeky when his mother introduced him to the wife of the new minister. The good lady asked Tony his age, and was quite taken aback at his reply.

'Twice my age, taken away from the square of my age,' the boy told her, 'gives a number which is seventy-seven more than Pam's age.' Grinning impishly, he added: 'And please don't bother about vulgar fractions.'

Tony's mother was most embarrassed, but she explained that Pam was the boy's younger sister, and of course the minister's wife wouldn't hold it against such a generous supporter of her husband's church.

But how old did all this make Tony?

72

It Fell From the Sky

They still can't agree as to which of them saw it first. But the object was certainly warm when they found it.

The boys claim it came from one of those flying saucers, and Jim even swears he saw the alien ship swoop down and up with a flash of flame before that lump of calcined rock crashed into the barn behind the farm. Of course it could have been a plain meteor, but why have things so simple?

Jack says it weighed seventeen ounces, but Joe's guess is twenty-six. John, who says he's figured it out from what he calls 'terminal velocity', insists it must have weighed twenty-one ounces. Jim's estimate is one ounce less than John's.

In fact, one of those estimates was only two ounces out; two of the other three boys were exactly equal in the amount by which their guesses differed from the correct weight.

Your guess is as good as mine, so what do you say?

73

What a Tie!

'Say, fellows,' cried Ben as his friend came into the crowded newsroom, 'take a load of his tie!'

'Yes, I know,' rejoined Bert, not at all put out by the laughter around him, 'but my wife chose it, and you know what that means.'

Ben nodded understandingly. 'So you had to waste at least three bucks! That's being married.'

'Cheap at the price,' chuckled Bert, 'but the tie didn't cost that much. It actually cost twice as many cents less than two dollars eighty as six would have cost in dollars.'

And even Ben kept quiet a few moments after that. But maybe you can say what the tie cost.

74
A Tale of Three Monkeys

'We'll share them out in the morning, boys,' grunted Flip. 'It's too dark now.' The three monkeys were tired. After toiling all day collecting a great supply of turtle-eggs, they were only too glad to curl up on the soft, warm sand for some welcome sleep. And so all became quiet, except for the murmur of the wavelets lapping along the beach.

But soon the rising moon wakened Flip. He peered at the other two and saw that they were sleeping. Edging cautiously to the pile of eggs, he broke one, swallowed its acrid contents, and scrabbled the shell into the sand. Carefully he counted the rest: carefully he counted out for himself exactly a third of them, which he hid deep in the sand where he again lay down and slept.

Time passed. Then Flap opened his eyes and saw the others fast asleep. He was hungry. Swallowing the slimy mess of one addled egg, he hid away just a third of the remainder. Then, well content, he fell asleep again in his place.

Soon there was a gentle stir and a rustle as Flop opened his eyes. Seeing the others deep in slumber, he gulped down an egg and then took for himself a third of the much reduced heap. And within minutes he was dreaming again.

And so when they wakened with the dawn, they shared the eggs which still remained, each taking an exact third and hoping his duplicity would not be suspected by the others. We don't know how they dealt with those hidden eggs, but we do know the three must have concealed between seventy and eighty eggs among them. Maybe you can figure out the exact number.

75

On the Queen's Highway

There was little traffic along the highway between Brandon and Portage la Prairie that day. But that was the day Ben drove from Portage la Prairie to Brandon, starting off at exactly the same time as his friend Bert; but Bert was driving in the opposite direction from Brandon.

They both drove at their favourite speeds—different, of course, but as near constant as makes no difference. When they passed each other, Bert had driven nine miles more than Ben, and required 64 minutes to complete his run to Portage la Prairie. Ben reached Brandon just 36 minutes after Bert reached his destination.

What do you make Ben's speed?

76

A Game of Canasta

'Let's not play for point-stakes today,' said Mary as the three friends sat down to play Canasta. 'I suggest the loser should double whatever money each of the other two has at the end of each game.'

Her two guests agreed with this novel idea—after all, Mary was hostess—and so they settled down to serious business.

Gwen's luck was right out at first: she barely managed to meld, and she lost the first game by a disgraceful margin. She made up for it in the next game, however, which she won.

Mary lost the second game. But the third was really exciting. It developed into a hard tussle between Mary and Gwen, each drawing lots of Wild Cards and going out several times on Concealed Hands.

After that third game, in which Clare had hardly scored any points at all, Mary proposed a break for tea. Pouring for her guests some minutes later, she remarked that she had won a little. 'I've got exactly sixteen dollars now,' she said. The other two checked their cash also and found they each had sixteen dollars too. It was a strange coincidence.

How much did Gwen have when they started to play?

77
Humpty Dumpty

Poor stumpty Humpty Dumpty! So loyal was he, so determined to see every detail of the great Royal Pageant. And now? 'All the Queen's horses and all the Queen's men—' well, you know how it goes.

Had he only lived three years longer, and been born (or is it 'laid'?) one third of his age at the time of his demise earlier, he would have died when he was as old as he would have been if he had lived to the age of five years less than twice the age at which he died.

How old was he when he had his fatal fall?

78
The Spreading Chestnut Tree

'That's a fine chestnut,' commented Ben, admiring the magnificent tree which towered in front of the farm-house. 'Yes,' replied the farmer. 'My dad planted it when he built this house. He used to tell me the sapling was eight feet tall when he put it in, and he said it grew exactly the same amount each year.'

'But d'you mean every year from the start?' asked Ben, who always liked to get his facts right.

'No, I wouldn't go as far as that,' admitted the old

farmer, 'but it was true for the first dozen or so years, and I remember being told that it was one-twelfth taller at the end of the ninth year than a year before.'

How tall do you figure that tree must have been at the end of the tenth year?

79

Only Snails

The two snails were hurrying along on their lawful occasions on a cylindrical column six feet around. Sammy was two feet up from the base at the moment when Simmy, on exactly the opposite side, was six feet up.

Just then some molluscan sixth sense made each aware of the other's proximity. Spring was in the air, and even snails have their moments.

'Which way will he come?' wondered Simmy, halting in her slimy slither. It was a long wait, but Sammy didn't disappoint her.

How far did he travel to her by the shortest route?

80

A Tale of the Cats

The lucky cats in Stratton Street
Had seven mice apiece to eat.
    The rest made do
    With only two:
    The total score
    Being twenty-four.
How many cats ate mousie meat?

81
A King Was Chosen

And now were the Elders assembled together in the Temple; and the sound of their voices was like unto the noise of thunder as each spake unto the other asking: 'Whom then shall we choose to be our King to rule over all Kalota?' And in that multitude there were many who would choose Kar the son of Knok; and there were some who called for Kapel: for it had been so ordained, and no other could be chosen, only of these two who now stood before the Elders.

Then came Keris the High Priest and spake, saying: 'All ye who choose Kar the son of Knok raise ye now the right hand.' And straightway were their right hands raised and they were like unto a field of corn before the reaping; and Keris went among the Elders and he counted; and two score and one of the Elders had not raised a hand; and eight score and five had raised only the right hand as Keris had so adjured them; and some, seeking advantage for their choosing, had raised two hands; and all the hands that were raised at that time numbered three more than the whole number of the Elders in that congregation.

And the High Priest went down from the high place and laid his hands upon Kar and blessed him and turned to the Elders, saying: 'Hear ye now all ye Elders of Kalota that are gathered together in this place from the farthest ends of the land, and let it be known by all men that ye have chosen Kar to be your King.'

That's what it says in the Chronicle of Kalota, but it would be interesting to figure out how many Elders voted for Kar.

82

Simon Bought Some Fruit

Simple Simon was sent out to buy some fruit, and came back with three bananas and half-a-dozen pears. When his mother asked what he had paid, however, the boy wasn't quite sure.

'I know the bananas cost one cent each more than the price of a pear,' he said, 'and I noticed that what I paid for the pears was the same as I paid for the bananas but reversed.'

His mother couldn't quite follow this: 'What do you mean by "reversed"?' she asked him.

'Well, I paid less than a dollar for each little lot,' replied Simon, 'and the amounts for the two lots were the same figures, but for one lot they were the other way round.'

This was even more confusing than before. But perhaps you can see how much Simon paid altogether.

83

Top Secret

The boys knew their father had been doing something rather mysterious during the war. He called it 'Intelligence', but then you'd hardly associate that with his having been so long up in Ottawa. Now he had just finished telling them about codes and ciphers, and was back to his newspaper.

But a seed had been sown. It wasn't long before Jack came over with a scrap of paper demanding attention. 'Add ode to ado to make deed', declaimed the boy. His father pondered a few moments, making some notes on the paper, and then he got it. 'That's a queer code,' he chuckled, 'using letters for figures.'

What was the addition sum?

84

In Kalota There Was a King

One hears more and more of the strange customs which make Kalota so interesting to visitors. But perhaps the strangest of all is the convention about speaking the truth.

The men never speak the truth, but never! And that really makes it very easy to know what they mean in conversation.

The women, on the other hand, always alternate truth and falsehood or vice versa, statement by statement. That does make things difficult but, compared with what happens in other countries, there's quite a lot to be said for this feminine vagary: a Kalotan woman does speak the truth half the time.

Now three of these delightful people were talking about their beloved King Kormon, who died some years ago: there seemed to be some doubt as to the year of his death.

'It was the year of Karan,' declared one of the three. 'You're wrong,' said the second Kalotan, 'it was the year of Kelkis.' The third Kalotan appeared to agree: 'Definitely the year of Kelkis.'

As a matter of fact their King Kormon did die in one of those two years, but unfortunately I've forgotten which. And, as I've also forgotten the sexes of those three Kalotans, I'm still trying to figure out the truth. But what do you make of it?

85

Mike's Pike

'What lovely fish!' exclaimed Mike's wife, admiring the three fine pike he had brought home. 'What do they weigh?'

'The largest weighs as much as the other two,' replied

Mike, 'the smallest weighs three pounds less than half the other two together, and the lot come to eighteen pounds.'

So what were the weights?

86

One Rainy Day

It was really too bad that it should rain today, of all days. The children had been so much looking forward to their big picnic, and now it was quite out of the question. The previous night they had gone to bed greatly excited and with high hopes, for the weather-man had said it would be fine and warm; then this morning they could only see rain and clouds.

But they had all assembled at the Browns' house as arranged: the Lamberts from across the river, the Beale boy from next door, and the Harvey kids from up the hill. And now the children were milling around in the recreation room, bored and disappointed, not knowing what to do with themselves.

Then Uncle Frank came to the rescue. Striding into the room, he slapped a quantity of dimes, nickels, and coppers on to the table. 'Here you are, kids,' he told them. 'Divide this equally among you; there are forty-five coins, but you'll have to fix change out of your own pockets.'

Sharing out the money, which came to $4.07 altogether, kept the children busy for quite a long time. Each received more than twenty-five cents; but can you figure out how many dimes there were? And how many children?

87

A Profitable Round

'What can I get you?' asked the storekeeper, seeing a well-known customer wandering rather aimlessly down the aisle. 'That depends on you,' was Clem's surprising reply. 'I'll buy ten dollars' worth if you will first double the money I have in my pocket.'

Clem is quite a character in the town, and he is always a good spender. The storekeeper decided to take a chance, and so the transaction was completed, and on a cash basis.

Leaving the store some time later, highly amused by all this, Clem decided to try the same routine in two other stores where they value his custom. As he had expected, the owners of those two were very ready to humour the whim of such a good customer even at the risk of some slight loss for themselves.

When Clem finally left the third store he found he had spent all his money and had not even a cent in his pocket, but his car was stacked with a varied assortment of merchandise. How much did he have when he entered the first store?

88

A Monkey on the Job

Today I met a monkey, but this was downtown where they are building that new store. He was dangling at the end of a rope, holding on with one paw and scratching happily with the other. Going closer, I saw that the rope passed over a pulley with a weight at the other end, and the two ends were exactly level.

This was all so strange that I made some inquiries from a man who seemed to be working there. Thus I learned that the monkey's age and the age of its mother total seven years; furthermore, the monkey weighs as many pounds as its mother is years old.

Intrigued by these revelations, I heard that the monkey's mother is one third again as old as the monkey would be if the monkey's mother were half as old as the monkey will be when the monkey is three times as old as the monkey's mother was when the monkey's mother was three times as old as the monkey was then.

This news staggered me, but more was to come: the weight of the weight and the rope came to twice the difference between the weight of the weight and the weight of the weight and the monkey.

I could take no more. 'Tell me the length of the rope,' I begged, 'and let me go.' My informant shook his head sadly: 'That we don't know,' he said, 'but four feet of it weigh one pound.'

I went on my way wondering, but still seek the answer to my question.

89

Three Friends and Their Birthdays

Christmas 1952 was over and the three teen-agers were looking forward to their birthdays. Bella was born in February, while Betty and Beryl have their birthdays the previous month.

Beryl is just eight weeks older than Bella, but three days younger than Betty. What year were they born?

90

It's Collars Today

'Well, this is a fine time to come in to work!' cried Sue, greeting her friend who had just entered the busy workroom. 'Why not?' replied Sal. 'That's the best of piecework.'

Sue nodded. 'We're on collars today,' she remarked, pointing to a little heap beside her, 'and I've made thirteen already. But you've got a different style, and your rate will be twice as much as I get, you lucky girl.' 'That's fine,' said Sal, as they both settled down to their machines. 'I reckon you'll do seven of yours to my four.'

For the rest of that day the two girls worked hard and fast, stopping only to have a quick cup of coffee together at the noon-hour. When they finally stopped that evening, it turned out that Sal's estimate was correct; and, strangely enough, her earnings for the day were exactly the same as Sue's.

How many collars did Sal make?

91

She Married Them Off

'What happened to those nice girls who used to run around with your Susan?' asked Mrs Jones who hadn't seen her friend for some years. Mrs Brown smiled: 'You mean Sally and Doris? They got married very soon after Sue.'

'Now isn't that nice. Such a relief for their mothers.' Mrs Jones was thinking of her own dowdy daughter who seemed destined for other things.

'Yes,' Mrs Brown went on, 'they married three old friends: Tim Morton, John Devlin, and Len Owens. Maybe you'll remember the boys.'

This was a good start for a real gossip, during the course of which Mrs Jones learned that Sal must be twenty-three; that Mrs Owens is one year older than Mrs Morton, and that Sue is two years younger than Doris. It also transpired that Mrs Devlin was as old as her husband was when he was three years younger than Mr Owens was when he was twice as old as his wife was when she was three years younger than Mr Morton was when he was half as old as his wife will be in seven years' time.

Mrs Jones came away slightly dazed, but perhaps you could tell her Susan's age.

92

The Beverage Room

Having heard so much about beverage rooms in Toronto, I entered one to see for myself. See? Well, not at first, but after a few anxious moments of ocular adjustment it did become just possible to see right across the room. What a shocking scene was then revealed! The place was full of people who seemed to have no regard at all for the niceties of the law.

Out of fifty-seven persons whom I counted there, no less than thirty-seven were smoking (not illegal as yet, but surely a pleasurable habit which must soon be banned). Forty-six sat there with elbow on table (shame on them for so relaxing in comfort). Eight actually wore glasses (disregarding the ban on 'two glasses at a time'). And among the throng there were forty-three persons who laughed as I watched them, contravening the spirit if not the letter of the law today.

But I had seen enough, more than enough. I escaped out into the freedom of the cool night air; that was at least untrammelled.

Having borne with me so far, perhaps you can figure out the least number of laughing, unbespectacled people I must have seen there, sitting with elbow on table smoking as they enjoyed their leisure.

93

Digging for Dollars

Bert and Bob were hired to dig a shallow trench from
Farmer Giles's house down to the pond. The trench was to
be 160 yards long, and the farmer told them to start at opposite ends and work towards each other: that way he
figured they wouldn't waste their time.

Bert, who started up by the house where the ground
was hard, was paid 25¢ a yard more than Bob. Bob was
quite content with this, as the ground was soft down by
the pond and he would be able to do far more than Bert.

When the job was finished, however, each received
exactly fifteen dollars. So how many yards did Bert dig?

94

More Files for the Office

'Will that be all?' asked the clerk. Larry toyed with the
stapler he had chosen. 'Oh, no. Mr Glen gave me twenty
bucks. He said to buy this first, and then spend the rest
on files, as near as possible to a hundred letter-size and
about fifty legal.'

The clerk reached up to a shelf behind the counter.
'How many of each did you say? Legal are twelve cents
each and letter-size seven.'

Larry was silent a moment, scribbling on a scrap of
paper. 'You'll have to help me figure this out,' he said at
last. 'Mr Glen's a funny guy. He hates having change
handed back; says it upsets the books. So I'll have to spend
twelve forty-seven on the files.'

'Certainly sounds cockeyed to me,' laughed the clerk,
trying to read Larry's figures upside down, 'but anything
to oblige a customer.' Needless to say, they did solve the
problem. But what do you make of it?

95

Best Irish Linen

Sal was showing Sue some of the Christmas presents she
had bought during their lunch hour. 'I got these for my
boy friend,' she said, opening a box of handkerchiefs.
'They're real Irish linen, but amazingly cheap.'

'How lovely,' exclaimed Sue. 'What did you have to
pay?'

Sal jotted down something on the back of an envelope
before replying. 'I told you they were a bargain,' she
smiled, 'and I bought one less than the price of each in
cents and spent three forty-two altogether.'

Sue shrugged her shoulders. It didn't really matter
anyway, as she had already finished her Christmas shopping. But what do you make the price of each handker-
chief?

96

Venice or Vladivostok?

The students at Exville High School are surprisingly
bright. A recent examination paper included two tricky
questions which thirty-seven of those students answered
correctly.

The questions were: 'Which is the more northerly,
Venice or Vladivostok?' and 'Which is later, sunset at the
Pacific or sunset at the Atlantic end of the Panama Canal?'

Of course some of them just didn't know. A third of
the students were wrong on the latitude question, but only
a quarter of them failed on the question of sunset. A
fifth of the students answered both questions incorrectly.

How many students were asked these questions?

97

How Old is Jack?

If Jack were two years younger than Jill would be if Jill were two years older than half as old as Jack would be if Jack were two years younger than twice as old as Jill would be if Jill were twice as old as Jack is, he would be ten years older than he is now.

All of this makes Jack's age quite clear—or does it?

98

Kurt's Birthday

'What day is Kurt's birthday next month?' asked Jim, one evening last June. Hilda smiled, happy to see he hadn't completely forgotten his 'in-laws' over in Europe. She probably didn't realize this was only one of the many little ways in which he strove to please her.

'You like problems,' she replied, 'so I'll tell you the way Kurt told me years ago. Add together all the dates of the days which follow his birthday in the month, and that'll give you three times what you get if you do the same for the days before his birthday.'

So what was Kurt's birthday?

99

The Story of Kalia

Now in Kalota there was a man and his name was Kurtis: and he grieved for that he had seven daughters but no son had been born unto him. And Kurtis went unto Kerin the High Priest to seek knowledge of what he must do.

And Kerin spake unto him saying: 'Be patient and the

Lord will bless thee and thy wife Kalia will bear thee a son.'
And Kerin spake also of the gold which Kurtis must give
as a thank-offering. And Kurtis swore that he would do
what was required of him, and he returned to his home
and did all that Kerin had enjoined upon him.

And it came to pass that Kalia bore him a son and
Kurtis rejoiced and his heart was full, and he went to the
Temple to give thanks and with an offering of gold.

And this was the count of his offering, even as he had
sworn unto Kerin the High Priest. For each whole year of
his age did he give five kalens of gold; for the first month
that Kalia his wife had dwelt with him did he take back
unto himself one hundredth part of a kalen of gold; for
the second month that Kalia had dwelt with him did he
take back two hundredth parts; for the third month three
hundredth parts; and in like manner for each whole
month thereafter that Kalia had dwelt with him.

And the gold which Kurtis gave unto Kerin the High
Priest was one score and three kalens, for he was a rich
man and thankful. And he lived in peace and happiness
with his wife Kalia and with his son and his daughters,
and was gathered to his fathers when he reached the age
of three score and ten years.

That's what the records say, but how old was he when
his son was born?

100

Three Sisters

Betty, Gwen, and Jane are sisters. Betty is as much
younger than one sister as she is older than the other.
Gwen is seven years younger than twice the age of Jane.
Jane is five years older than half the age of one of her
sisters.

How old does that make Gwen?

101

Spot the Winner

Jim and Joe were arguing about last year's athletics, and in particular about the 100-yard sprint for which there had been only three runners.

'But I tell you I saw the finish,' insisted Joe, 'and Ken won the race with Steve a close second.'

But Jim shook his head. 'You may have seen it,' he told his friend, 'but you've got it all wrong. Ken was second and Larry first.'

And neither of them was right, as each had made one mistake regarding the final placing of those three runners. But perhaps you can spot the winner.

102

They Played Checkers

Mike and Steve had just finished their sixth or seventh game of checkers when Mary came over to the table. 'No more of that, Steve,' she said. 'You can be sociable now for a change, and that goes for Mike too.' Steve looked up with a grin, knowing better than to argue with his wife: 'That's all right by me, but I've won a bit and I wanted to give Mike a break.'

Betty had come over to join them and was ruffling her husband's hair. 'Sorry, Steve,' she laughed, 'but you'll have to give him his break another day. We're tired of sitting over there alone, whispering for fear of disturbing the game.' She picked up some coins which lay on the table beside Mike. 'So you stop right now, but tell me how much you've lost.'

'Only five fifty-five altogether,' replied her husband obediently, 'but we played for unusual stakes. Each game we played for half of what I had in my pocket at the time, and we paid up after every game.'

He could not say how much he still had left, for Betty had put most of it into her bag. Maybe you can figure out the amount.

103

The Travel Set

'Thank you, darling!' Mary squeezed her husband's arm. 'Now I really do feel I'm going places.' She stroked the smooth cowhide of the beautiful wardrobe case, wishing Steve could go with her on the trip out west.

'You needed them,' he told her, 'and they were a bargain anyway at five dollars off the regular price.'

Steve had just bought her the handsome two-piece travel set. Assuming the overnight case was worth two-fifths of the regular price for the set, he figured the wardrobe case had only cost him eleven dollars seventy-seven.

What did he pay for the set?

104

A One, a Four, and a Nine

Take a 'one', a 'four', and a 'nine'—all three of them, but only one of each. Then, using any regular mathematical signs which you may require, form an expression for the number 1.

That won't be difficult, but now you have to do the same for every number in turn, working upwards as far as you can go.

As a simple example, you could express 4 as $(9 - 4 - 1)$, but then there aren't many as easy as that: you'll certainly run into difficulties by the time you reach the twenties.

105

Shoes for Free

'Bad news for you, Bert,' said Bob, as he entered his neighbour's shoe-store. 'That twenty-dollar bill I changed for you yesterday was a fake. The bank just told me.'

The previous day a complete stranger had bought a pair of shoes for $17.00, paying with a $20 bill; Bert, finding himself short of change, had slipped round to his friend's shop to get the bill changed, and so had been able to give his customer the $3 due to him.

'Well, for heaven's sake! I'd have sworn that bill was genuine,' said Bert, more horrified by his own stupidity than by the actual loss. He then explained the circumstances while counting out twenty one-dollar bills for his friend. 'Anyway, *you* don't have to lose anything,' he told Bob, handing over the money.

What was the true extent of Bert's loss?

106

That Was a Family

Ron is very attached to his aged grandfather, and really enjoys his stories of the past. When the old man embarks on a long rambling tale of his boyhood, however, it is sometimes difficult for Ron to keep track of all the brothers and sisters whose names are mentioned.

During one such story Ron interrupted his grandfather. 'How many brothers and sisters did you have?' he asked.

The old man went on with his tale for a moment, ignoring the question. And then he stopped, peering at the boy over the top of his glasses. 'What was that? How many of us?' There was a merry twinkle in those tired old eyes: 'You're a bright lad, so you can figure it out yourself. My sister Prudence had three times as many brothers as sisters, while I had only twice as many.'

Ron did figure it out. But what do you make of it?

107

John and His Bicycle

'You'll have to hurry, dear, if you want to get your bike today,' said his mother, and so John set out for the village, walking at a steady four miles an hour.

When he reached the repair shop the bike was ready and he only stayed five minutes there before starting back home. It's a good road and he cycled back at an average speed of twelve miles an hour, spurred by thoughts of the supper which would be waiting for him.

'See, Mum,' cried John, as he entered the house, 'I've only been away five minutes over the hour.'

How far is the repairer from John's home?

108

Who Broke the Cups?

It's always the cups that get broken, and Mary has such an accumulation of odd saucers. So today she made a point of buying some extra cups; she chose good china, but plain white, as that would be easier to replace.

The saucers cost 19¢ each but she had to pay 37¢ apiece for the cups, and you'd say she got good value for her outlay of $5.58. But can you say how many saucers she bought?

109

Findings isn't Keepings

'Look, Dad!' cried Ken, rushing into the living-room with a muddy leather purse in his hand. 'See what I just found in the road!' 'You'll have to take it to the police,' his father told him.

But curiosity is a very human failing, and so Ken's father added: 'What's inside?'

'Only a few coins,' the boy replied, 'quarters, dimes, and nickels, but I didn't count them.'

His father took the purse and checked its contents. 'I don't expect they'll give you anything, so I'll give you your reward myself,' he said, 'but you'll have to tell me how much there is here.' He checked again, and went on: 'The dimes and nickels together are one more than the number of quarters; the quarters and nickels are three more than the dimes; and the dimes and quarters five more than the nickels.'

How much would you say the purse contained?

110

Not for Children

'Well, dear, you must do what you think best, but it sure is a problem,' said Steve to his wife as Garry came into the room.

The boy was quite accustomed to this sort of thing; so many of his parents' conversations seemed to fizzle out at his appearance. But this time he'd caught the word 'problem'.

'Here's one for you, Dad,' he said. 'When Jane was as old as the number of years Jack is older than John, she was half as old as Jack was when he was half as old as John is now. The ages of the two boys add up to eighteen years, so how old is John?'

What would you say?

111

Which Road for Knokado?

You will remember that Kalotan women never make two
consecutive true or false statements: if one statement is
true then the next is false, and vice versa. This is one of
the peculiarities of those fascinating people, and it certainly adds to the excitement and uncertainty of life in Kalota.

Over there last year on a visit, I had an interesting ex-
perience which illustrates how this feminine vagary can
be used to overcome the very difficulty it creates. I was
driving to Knokado, the capital of the island, and came to
a spot where the road divided: one way led right and the
other left, but there was no sign or other indication as
to the proper road.

A peasant woman was standing there on the roadside,
so I asked her. Before doing so, however, I recalled where
I was: in other countries women can never be trusted,
but this was Kalota. So I asked just one question: one
question worded most carefully.

The woman replied quite civilly. And then, realizing
that she must surely know the way to the capital, I continued my journey in absolute confidence and so arrived
safely at my destination. Later, as a matter of interest,
I did find out that the other road would have taken me in
quite the wrong direction.

It may amuse you to figure out the wording of my
question.

112

Sam Rides Home by Taxi

'Look! There's Sam, just got home by taxi,' Mary pointed
across the road. 'He always comes back that way now.'
'Yes, I know,' replied her husband. 'The poor guy had his
licence cancelled.'

But Mary has a horror of extravagance. 'Why doesn't he use the bus and subway?' she persisted. So Dan had to do his best to explain Sam's reasoning as his friend had put it to him a few days before.

Living quite a way from his job, Sam had come to the conclusion that he would be wasting valuable time coming home by public transport; and there was also the detail of personal comfort to consider. By subway and bus the journey cost only 27¢; but a taxi, although charging a quarter for the first fifth of a mile and a nickel for each fifth after that, saved him more than an hour.

Working additional overtime at the factory, and completing the necessary number of shirts at the piece-work rate of 28¢, Sam figured he just covered the extra cost of the taxi each evening. He had mentioned that the ride happened to be an exact number of miles, but Dan wasn't sure what this had to do with it.

Mary listened carefully to all this, but she wasn't at all convinced. 'What about the tip, and his income tax?' she objected. Dan laughed: 'Well, I guess he didn't think of that, but anyway don't mention that to his wife.'

What do you make that taxi fare?

113

For Betty's Birthday

'What's on your mind?' asked Mr Kemp. Joe coughed nervously: 'Well, sir, it's Betty, your secretary. We're planning to get something for her birthday, and the boys thought you might like to subscribe.'

The manager smiled. 'Only too glad,' he said, 'and I'll give six dollars more than a fifth of what you've collected so far.'

'Gee, but you're generous. Nobody's given more than three bucks,' exclaimed Joe. He glanced at a slip of paper in his hand. 'That'll increase the average sub by exactly a dollar.'

How many had subscribed already when Joe saw Mr Kemp?

114

Only One Girl Among Them

Bill, Joe, and Sam are neighbours. Their surnames are Leary, Yates, and Renoir, but not necessarily respectively. Each has a grown-up son, but among their six children there is only one girl.

Yates has one child less than Sam, while Joe has as many children as Sam and Bill together. Renoir has the same number of sons as Yates.

So what is Leary's first name, and how many sons has he?

115

Their Birthday

It's their birthday, you see,
The same for all three,
It's strange but it's perfectly true.

There's Bertie and Ben,
Who differ by ten;
Eight years older than one there is Sue.

Double one brother,
Plus treble the other,
Plus Sue's age makes seventy-two.

From what's on this page
You can find the girl's age.
It's really quite easy to do.

116

The Flatterer

'Good morning, and how are you today?' Sally greeted her neighbour as they met in the store.

'Not too good,' replied Amelia in resounding tones which belied her words. 'If you were as old as I feel, and if I had been as young ten years ago as you look today, you'd be twenty years older than me.'

Before Sally could quite determine whether this was unexpected flattery or much the reverse, Amelia went on: 'And really, my dear, you look twenty years younger than I am.'

So how much older did Amelia suggest she felt that day?

117

That Jack Again

'How old's Aunt Emily?' asked Jack, looking at the snapshots which had come in the mail.

'You can figure it out yourself,' his father told him. 'She's thirteen years older than I was when Aunt Claire was half my present age. And she is five years older than Aunt Claire, and two years younger than me.'

Maybe Jack knew Aunt Claire's age, but anyway you can figure out how old Aunt Emily was.

118

The Telephone Number

'Is Mr Brent in?' asked Ken. 'I'm sorry, sir, but he's out and we don't expect him back today. May I tell him who called?' The telephone operator sounded quite charming and most helpful, obviously a new girl. But Ken could not wait until the morrow. 'It's Ken Keeling,' he told her. 'Maybe you can give me a number where I can call him.'

For a moment there was silence apart from a faint sound of rustling paper, and then she was back. 'Mr Brent did leave a message for you. If you double the number where he'll be and subtract your age, he said you'll get the exact reverse of that number on the Park exchange.'

Ken swore to himself, but thanked the operator before hanging up. 'Curse Sam's sense of humour!' he muttered. But this was urgent, so he could only sit down and waste precious minutes figuring it out.

Of course you'll have to know that Ken is thirty-nine, and then you'll surely find the number.

119
Betty's Birthday

Last week John and Betty had their birthday. They aren't twins, although they do look so much alike, but it happens that Betty has the same birthday as her elder brother. And this year he noticed something rather special about their ages.

'D'you know?' he said. 'If you multiply our ages together and then add your age and mine, you get the number of Dad's car.'

Betty considered this a moment. 'What, you get 142?' she asked. John nodded. 'Sure, that's what I'm telling you.'

But the figuring was rather beyond Betty, and anyway she couldn't see that it mattered very much. John was right, however, so how old does that make Betty?

120

Simple Simon Again

Simple Simon was sent to the dairy to return some empties and buy milk for the weekend. He returned home with seven bottles of Jersey for which, he told his mother, he had paid 29¢ each. 'That included the bottle,' he added, 'and they allowed me the usual nickel on each of the empties.'

But when it came to handing over the change from the money he'd been given, Simon ran into trouble. He'd mixed it up with his own cash, and of course he couldn't remember how much he'd paid, and nobody seemed to know how many empties had been returned.

'I know I gave them between a dollar and a dollar fifty,' said the boy, 'and I did notice that the amount was exactly divisible by seven.'

What would you do with a son like that? His father, who had been listening to all this, sighed and shook his head. 'Let's hope he'll grow out of it,' he told his wife.

Can you figure out how much cash Simon spent?

121

A Parcel From Each

Jack and Jill are becoming quite good at figuring out the peculiar problems which result from their mother's inability to keep track of what she spends shopping. Yesterday Sally Brown came home with five parcels, having spent all her money; and she wanted to write down what she had spent in her little black book.

'Now you figure it out,' she told the children. 'I went into only five stores, and in each I remember I spent a dime more than half what I had in my bag when I entered.'

Jill finished first, but said nothing until Jack was through; and they agreed on the amount. So how much did Mrs Brown spend?

122

That Old, Old Story

The old riddle starts with the host pointing to a portrait on the wall. In the modern version, which starts the same way, he might go on to say: 'Brothers and sisters have I none, but that man's father's a score and one years older than that old son-of-a-gun, whose father of course is my father's son.' Pausing a moment, he would continue: 'Forty was he when his portrait was done, two years ago by an artist for fun.'

Now, how old would the speaker be?

123

A Glad Goodbye

Ken was glad to see the last of Aunt Amelia. He hadn't enjoyed her visit at all. All day long it had been 'Don't do that' and 'You mustn't do this'. She was worse than Miss Prym at school.

He heaved a happy sigh as he climbed back into the car after seeing the train vanish round the bend. 'I'm sure glad the old hag's gone!'

His mother was shocked. 'Ken,' she scolded, 'don't you dare talk about my sister like that. And anyway she's younger than me.'

The boy saw he had gone too far this time. 'I'm sorry, Mum,' he told her, 'but how old is she really?'

Ken's father secretly shared his son's sentiments, but

he answered the question. 'Three years ago,' he said, 'she was five times as old as the difference between Clem's and Clare's ages, and in a year's time her age will be twice their ages added together.'

'But who are they?' asked Ken, never having heard of Clare or Clem. 'Just two of your many cousins, son,' chuckled his father, 'and one's a teenager.'

So how old does that make Aunt Amelia?

124

Some Christmas Shopping

'You wait out here,' Mary told her husband as she turned back to re-enter the big store. 'I won't be long.' Steve was only too glad to miss another battle with the hordes of Christmas shoppers who thronged in the aisles inside. 'But won't you need some more cash?' he asked.

Mary checked in her bag. 'I've got plenty, dear,' she assured him. 'It's only one little thing I forgot. I spent a quarter of my money in there already, but I've still got a dollar more than I would have if I'd spent a third instead of a quarter.'

That gave Steve something to think about while he waited. How much would you say she still had in her bag?

125

The Commuter

The Clarkes are punctual people. They live well away from Vancouver where Mr Clarke works, and they don't envy city folk. He commutes, and returns home every evening by the same train which, strange to say, always runs exactly on time. And his wife meets him at the station, leaving home at the same time every day and driving always at the same steady speed to reach the station simultaneously with the train. It's an unusually well-ordered life they lead, as you can see, but they are very happy.

One day recently, however, things went wrong at home. Mrs Clarke is never one for excuses so we won't know the details, but she certainly left the house thirty minutes late.

On getting off the train, Mr Clarke realized that something very much out of the ordinary must have delayed his wife, and so he didn't waste time at the station but set out on foot.

Just as he was beginning to weary, his fond wife appeared, bundled him into the car, and drove back home. They entered the house just as the old grandfather clock struck six. They were twenty-two minutes late.

Assuming Mrs Clarke drove at her regular speed that day, despite the circumstances, how long had her husband walked before she picked him up?

126

One Fake Dollar

A collector had eight Edward VII silver dollars, all identical and in mint condition. Showing them to some friends one day, he also showed them a counterfeit coin which appeared to be identical with the real dollars, but he told them that the imitation was slightly lighter than the genuine article.

Unfortunately the coins became mixed while being examined by the guests. The collector owned an accurate balance-scale, but had no weights for it. Nevertheless, using only this, and in only two weighings, he found the fake dollar.

Can you figure out how he did it?

127

Fun in the Tub

Junior must have been in the bathroom at least half an hour. When he finally emerged, his mother was waiting outside all set to give him a good scolding.

But the boy forestalled her. He explained he had been figuring out some school work while in the tub. 'It's the tiled wall,' he told her. 'If each tile were an inch smaller there'd have to be six more tiles to the row along that wall: but if each tile were an inch bigger there'd be four tiles less.'

His mother had little patience for such nonsense and she packed him off to bed. But how long was that wall?

128

Only Partly Right

The three boys had almost come to blows when their father burst into the recreation room to see what all the noise was about. They told him they were arguing about their birthdays last year.

Tam contended that Tim's birthday had been on a Monday and Tom's on a Friday.

'Mine wasn't Monday!' shouted Tim. 'But yours was a Tuesday.'

Tom disagreed with Tim about the Tuesday but maintained that his own birthday was on a Monday.

Their dad laughed. 'Those were the three days, sure enough, but you may be interested to know,' he told the boys, 'that each of you has made one true statement and one false statement.' Making for the door, he added: 'So now you can figure it out yourselves, and let's have no more noise.'

What days were their respective birthdays?

129

Neighbours

Fergus and Fred lived on the 'odd-number' side of their road. Sitting in his friend's garden one day, Fergus remarked that the number of his own house, added to a hundred times the number of Fred's house, gave the four-figure number of the latter's car.

'That's funny,' replied Fred, 'but what about this? Our house numbers, multiplied together, also give the number of my car.'

That is what happened, but you will have guessed what this is all leading to. You have to know that the first figure of Fred's car number was not a nought, and then you can figure out what that number was.

130

The Ballad of Ballygan

To his dismay
He learned next day
What havoc war had wrought:
He had at most
But half his host,
Plus ten times five two nought.

One-ninth were lain
In beds of pain,
With hundreds eight beside;
One-sixth were dead,
Enslaved or fled,
All lost in warfare's tide.

And so 'tis told
To warn the bold:
Aggression doesn't pay.
How many men
Marched with him, then,
At dawn the previous day?

131

His Dinner Waited

A fat and foolish fly was fast asleep just one foot down
from the ceiling halfway across the end wall of a room,
thirty feet long from end to end. Halfway across the
opposite wall there prowled a hungry spider in search
of food.

His senses sharpened by appetite, the spider suddenly
became aware of the sleeping fly at the very moment when
he was one foot up from the floor. In eager anticipation
of the feast, he scurried towards his prey by the shortest
possible route.

The room, bare of windows and boasting only one
narrow door, was twelve feet high and twelve feet wide.
It was cold and cheerless, but the spider had a date with
his fly. How far did he travel to reach his sleeping victim?

132

How Young is Young?

'Saw you with a good-looking young man at the Mem-
orial Arena last night,' remarked Peg as she joined her
new friend at the street-car stop.

'Oh, that was only my brother,' replied Pam.

But Peg seemed to have been impressed. 'I'd like to

meet him some time,' she suggested.

Pam laughed. 'He's much too young for you, much younger than you'd think. When he's as old as I shall be five years before I'm twice as old as he is now, he'll be sixteen years older than I was when I was half his present age.'

Peg looked at her open-mouthed. She had never expected an answer like that. But can you figure out the boy's age?

133

The Art of Selling

Simple Simon's sister Sarah worked in a cigar-store for a few days after leaving school. A little book on the magazine stand told her all she wanted to know about selling, and she faced the first customer with confidence.

'How much are those?' asked the man, pointing to some gaily-banded cigars.

Sarah didn't hesitate. 'A cigar and a half per day and a half for a week and a half would cost a dollar and a half,' she told him. That's what the book had said: 'Be different. Make them interested.'

And it worked, for he laughed and bought a dollar's worth. But how many cigars was that?

134

How High Was the Window?

It takes a youngster to notice things. Out with his Dad, Ken suddenly pointed across the street. 'Look! There's something hanging down the front of that house.'

They crossed over to investigate and, sure enough, a

fishing-line dangled down from a closed upper window. At the end of the line was a little lead sinker which just touched the paving of the sidewalk.

'Now, Junior,' said his father, 'we'll find out how high that window ledge is.' Taking a steel measuring-tape from his pocket, he pulled the sinker to one side until it was two feet above the paving, keeping the line taut. In this position the sinker had been displaced horizontally exactly ten feet.

By this time Ken had lost interest in the whole affair, but he made a good show of pretending to understand the way his father figured out that height. How high do you make it?

135

First Snow of the Winter

It was early November. 'Cold, with snow,' the weatherman had predicted, and this time he was right. From Bloor right down to College Street there was a complete hold-up of street-cars—some trouble with a teenage driver who had side-swiped a truck with his ancient jalopy, shedding a wheel or two in the process.

'We'll get out,' Mike told his wife. 'It's much quicker walking.'

Sliding and slithering through the slush, Betty followed slowly in his track as they struggled up Bay Street. For every pace forward, Mike slipped back a third of a pace; Betty, faring better with her shorter steps, slipped back only a quarter of every step forward. But Betty took five steps while Mike took three.

When Mike reached Wellesley Street he was surprised to find his wife still close on his heels. 'How are you doing?' he shouted through the driving snow. 'Fine!' was her cheery reply. 'But you just keep on bull-dozing.'

Mike's forward strides averaged thirty inches. How long were Betty's?

136

The Expense Account

Bert and Ben were discussing their jobs. Although they sell for rival firms, they still remain the best of friends.

Today Bert was thoroughly disgruntled. 'I've had another row with them about expenses,' he complained, 'and it's always the same. They're just mean.'

Ben smiled rather patronizingly. 'You haven't trained them right,' he told his friend. 'Now there's never any trouble with my people,' he went on. 'Last week I spent as much on Monday as on Wednesday, on Tuesday two bucks more than the amount by which Thursday's expenses were less than Monday's, on Thursday five dollars less than on Friday, and on Wednesday I spent more than six bucks.'

Bert was completely lost. 'Gee,' he gasped, 'that was sure some expense account.' But Ben interrupted him: 'And the funny thing was that each day I spent an exact number of dollars, and the total was twenty-nine dollars for the week.'

Presumably Ben doesn't make out his account in such an obscure form, but anyway it would be interesting to know what he spent on Friday.

137

Sal and the Skirts

Sal and Sue are still employed as operators in the factory. Just recently they have been working on a special line of women's suits, Sue on the coats and Sal on the skirts. It was most monotonous work—eight hours a day, day after day, turning out garments with almost automatic regularity: always the same design, the same material, and even the same size fitting. But the money was good, and the girls didn't complain.

Sal was making twelve skirts more in a day than the number of coats Sue made. The time it took Sue to make the number of coats by which the number of skirts Sal made in a day exceeded the number of coats Sue made in two days was the same as the time it took Sal to make two skirts fewer than the number of coats Sue made in a day.

So now we know how many skirts Sal made in a day, but do you?

138

The Five Threes

In this skeleton sum, two 3-figure numbers are multiplied together. There are only five threes, however, and these are shown; all the other figures are indicated by crosses. Now we have to figure out the two numbers and their product.

$$
\begin{array}{r}
x\,x\,3 \\
x\,x\,3 \\
\hline
3\,x\,x \\
x\,3\,x \\
x\,x\,3 \\
\hline
x\,x\,x\,x\,x
\end{array}
$$

139

A Trip to Neepawa

Ken was tired of sitting in the back of the car, watching his father's neck and wondering why it wasn't smooth like Mum's. 'What sort of a town is it?' he asked. 'Will

I be able to see a show tomorrow?'

'Neepawa's bigger than Roblin, where Uncle Pierre lives,' replied his father, 'but we'll have to see what Aunt Claire has fixed for tomorrow.'

But Ken was still curious: 'How much bigger?' he persisted. His father thought for a moment. 'This'll keep you busy for a bit,' he chuckled. 'If two hundred and fifty-four people moved from Neepawa to Roblin, Roblin would have half the population of Neepawa; but when we three arrive in Neepawa the population will become three times that of Roblin. And just forget about other visitors,' he added, as an afterthought.

For a long time there was only the sound of muttering from the back of the car as Ken scribbled away, but the boy figured it out. What do you make the population of Neepawa at that time?

140
The Curse of Man

Bert burst out of the bathroom, blood on his chin, swearing he'd give up shaving. 'You'll do nothing of the sort,' laughed his wife, 'for you know what I think of beards. And besides, it's your birthday, so stop cussing.'

Sitting down at breakfast some minutes later, Bert came back to the subject. 'I've shaved every day of my life since Dad gave me my first razor on my eighteenth birthday,' he remarked, 'and if I had shaved for six times as many months as I have years and for three times as many weeks as I have months I should be exactly five years older than I am today.'

This was not quite accurate, for he had forgotten about leap years, but maybe you can figure out his age.

141

Peace at a Price

'All right, I'll play with you, but only for a few minutes,' said Uncle Jim, knowing very well his nieces would give him no peace unless he did so. He jingled a fistful of coins and promised them to the one who guessed their total value.

'A dollar sixty-five,' hazarded Pam. After some more jingling, Pat took a chance. 'Two forty,' she suggested. Peg was quite confident. 'Dollar fifty,' she cried, 'so now pay up and no cheating.'

But Uncle Jim opened his hand and they saw they were all wrong. He saw his chance to buy peace, however. 'Pam was nearest,' he said, 'so I'll give her the prize if you'll all leave me alone for the rest of the afternoon.'

In fact, one of their guesses was twice as much in error as one of the other two, and that in turn was twice as much in error as the third. So how much did Uncle Jim have in his hand?

142

Doreen Got Her Puppy

'Isn't he cute?' exclaimed Doreen, pointing to the spaniel pup in the window of the pet shop. 'I want him, Daddy.'

Her father, indulgent as fathers are where small daughters are concerned, allowed himself to be led inside. They inspected the very mixed and somewhat dejected collection of cats and dogs and birds, but Doreen was still set on the spaniel.

'Okay,' laughed her father, 'there are thirty-six animals here including the birds, and you can have your puppy if you tell me how many legs there are altogether.'

Doreen counted frantically and arrived at a total of

exactly one hundred. She was far too young to even think of classing a human being as an animal, and so she got the pup. How many birds would you say there were in the shop?

143

So Many Bananas

'What a lot of bananas!' exclaimed their mother when the children laid out their purchases on the big kitchen table. She turned to Bill, older and perhaps more responsible than his two sisters. 'What did they cost?'

'The girls bought them,' replied Bill. 'Jane got some and Mary got some also, but at the same price, and altogether they spent a dollar eighty.' He hesitated a moment and then added: 'If Jane had paid as many cents each as the number that Mary bought, what she got would have cost ninety-six cents.'

'And each would have cost me an exact number of cents more than they did cost,' said Jane.

'You mean "it", not "they",' laughed her mother, but she was far too confused to count those bananas.

How many did the girls buy altogether?

144

On the Road

'Another day off!' exclaimed Mike, seeing the fishing equipment in his friend's car. 'You're the lucky guy.'

'But I make up for it in the winter,' replied Steve who covered a wide territory as a textile salesman, 'and anyway I'll be working harder next month.' He paused a moment, and then went on: 'Last month I averaged twenty-three dollars for each day I was on the road, but for two of those days I made only twenty-seven bucks altogether.'

Mike knows his friend's movements very well, and he certainly knows how many days Steve had worked the previous month. 'Apart from those two bad days, then, you must have averaged twenty-four dollars for the other days you worked,' he commented enviously.

How much did Steve earn that month?

145

A Lady and Her Lingerie

His lady was lovely, with glamour galore.
She led him one day to a lingerie store.
A dozen of these and two dozen of those
Transparently feminine silk so-and-so's
Would cost him, they said, six bucks less, if you please,
Than a dozen of those and two dozen of these.
Such extravagance, though, was way out of his reach.
He'd only six bucks, which just bought one of each.
Ignoring the frowns and 'tut-tuts' of the grundies,
Just figure the price of each item of undies.

146

Her Secret

'How old are you, Gran?' asked Ken politely. The old lady smiled at the little boy. 'That's a question no gentleman should ask,' she replied, 'but I'll tell you if you promise to keep it a secret.'

Ken nodded, and his grandmother went on: 'If you

reverse my age you'll get half of what I shall be in a year's time.'

That kept him quiet for quite a while, but what do you make her age?

147

The Marbles Rolled

If you want to reduce that waistline, just drop fifteen marbles on the floor of a swaying street-car and then try to recover them without bending your knees. That's what Garry did, but he bent his knees—he has no waistline worry, anyway.

He had been admiring his treasures: red marbles, green marbles, blue marbles; seven of one shade, five of another, and three of a third colour. Then the car gave a violent lurch and the whole lot went skittering over the floor. But Garry managed to recover all but four of them before he had to alight with his mother at their stop.

'How many of those pretty greens have you left?' asked his mother, genuinely sympathetic. 'I've got two more reds than greens,' he replied rather sulkily, 'but of one colour I've now only got two.'

Now, how many green marbles did he have before the street-car lurched?

148

Late, as Usual

'You're late,' called out Betty from the front porch. 'Steve left some time ago. You know how he hates speeding, but he said you'd overtake him.'

Mike looked at his watch. 'Yes, I'll sure do that long

before he gets to Polmood. If I averaged only thirty-two
I'd catch him in three hours, and at thirty-six in only
two.' Steve is one of those cautious drivers: he dodders
along at what he calls his 'safe speed'. Mike knows the
exact figure, and laughs at his friend—'As if any one
speed was safe!'

However, Mike wasn't wasting any time today. He
put his foot down and drove off in a cloud of dust. With
a fairly clear road all the way he was able to maintain a
steady forty miles an hour as he had expected. So how
long did it take him to overtake Steve?

149
But How Old Was Pam?

'Say, Dad, how old is Mum?' asked Ken. 'Jack says she's
older than his mother, but Mum looks years younger to
me.'

'That's right, son,' agreed his father, 'but don't tell Jack
I said so.'

Mr Dale thought for a moment, and then continued:
'You can figure out your mother's age yourself. Add together the two figures of her age, and multiply by the
age of your younger sister, and you'll get what your Mum
was a year ago.'

Ken's younger sister, Pam, certainly cannot be less
than seven, but we'll leave you to figure it out.

150
Cookies for Four

Johnnie was sent out to buy some cookies. 'You'll have
to give me half the cookies and also half a cookie,' his
mother told him, 'and then you'll have to give Agnes half

the remaining cookies and also half a cookie, and after that Jack will get half the balance and also half a cookie.'

'Okay, Mum, but what do I get?' asked the boy after he'd repeated the rigmarole a few times to himself.

'You know you don't like cookies,' he was told, 'but fix it so that there'll be just one over for yourself.'

And that's just what he did. But how many cookies did he have to buy?

Typical Solutions

2. Cigars for Dad

Cheap cigars cost x¢ each, so he had $(16x - 8)$¢.
Medium-price cigars cost $2x$¢ or $(x + 6)$¢ each. **Hence:**
$$16x - 8 = 24x \text{ or } 12x + 72$$
whence $\quad\quad\quad\quad x = -1 \text{ or } 20$
Cigars could not cost a negative amount, so $x = 20$.
Hence, the prices were 20¢, 26¢, and 40¢.
He spent $3.12.

1. Murder on the River

He paddled at x miles per hour through the water.
The current was y miles per hour over the land.

After 15 minutes he was $\dfrac{x}{4}$ miles from the glove.

The glove took $\dfrac{1}{y}$ hours to reach steps.

So after he turned, the glove took $\left(\dfrac{1}{y} - \dfrac{1}{4}\right)$ hours

i.e. $\quad\dfrac{4 - y}{4y}$ hours to reach steps.

In that time he moved through the water $\dfrac{x(4-y)}{4y}$ miles.

But we have seen he was $\dfrac{x}{4}$ miles from the glove.

So $$\frac{x}{4} = \frac{x(4-y)}{4y}$$

i.e. $$1 = \frac{4-y}{y}$$

whence $$y = 2$$

So the speed of the current was 2 miles per hour.

TYPICAL SOLUTION C

3. How Old is 'Old'?

As the maximum error was 9, and the difference between lowest and highest estimates was 12, the correct age was between 27 and 39.

Say the error of 27 was a years and the error of 39 was b years, then

$$a + b = 12$$

Errors were 1, 3, 6, 9.

Hence we must have:

$$a = 3, b = 9 \text{ or } a = 9, b = 3$$

with the correct age 30 or 36

But 36 was one of the incorrect estimates, so the correct age was 30 years.

TYPICAL SOLUTION D

4. The Spider and the Fly

Flies have 6 legs.

The spider would have required x females with 8 legs, y males with 3 legs.

Hence $$8x + 3y = 28$$

This is the simplest form of 'indeterminate equation'—
one equation with two unknowns.

Divide through by 3 (i.e. the smaller coefficient).

Then $$3x - \frac{x}{3} + y = 9 + \frac{1}{3}$$

Now x and y are whole numbers and therefore $\dfrac{x+1}{3}$
must be a whole number.

Say $\dfrac{x+1}{3} = t$ where t is any whole number, positive
or negative or zero.

Then $$x = 3t - 1$$

Substitute this value for x in the original equation, so
getting $$y = 12 - 8t$$

In this particular case we know that x and y must both
be positive, and so we see that t must be 1.

This gives $$x = 2, y = 4.$$

Hence there would have to be 6 'flies like that': 4 males
and 2 females.

TYPICAL SOLUTION E

5. A Tale of the Subway

The ages are: Mum, x years; Bob, y years.

From data given, we have the equation:
$$5y^2 - 2xy + 3x + 6 = 0$$

This is an 'indeterminate equation of the second degree':
two unknowns, including the square of one or both. For
the first step we treat it as an ordinary quadratic, in this
case in y:

$$y = \frac{x \pm \sqrt{(x^2 - 15x - 30)}}{5}$$

As x and y are whole numbers, the expression under the
square-root sign must be a square, and positive.

Now let $$x^2 - 15x - 30 = k^2$$

where k is any whole-number that will satisfy the 'square'
condition.

Treating the equation $x^2 - 15x - 30 - k^2 = 0$ as a
regular quadratic, we get:

$$x = \frac{15 \pm \sqrt{(4k^2 + 345)}}{2}$$

Again, the expression under the square-root sign must be a square, and we have to find values of k that will make it so. There are various ways of dealing with this, one being by trying successive values. In this case there is an easier method.

$4k^2$ is a square, which simplifies matters.

Let $4k^2 + 345 = t^2$ where t is a whole number.

Then $\qquad\qquad\qquad t^2 - 4k^2 = 345$

i.e. $\qquad\qquad\quad (t + 2k)(t - 2k) = 345$

Taking the factors of 345, we have:

$345 = 345 \times 1$, or 115×3, or 69×5, or 23×15. Now tabulate these alternatives, coupled with the expressions $(t + 2k)$ and $(t - 2k)$, where $(t + 2k)$ must be greater than $(t - 2k)$:

$$t + 2k = 345 \text{ or } 115 \text{ or } 69 \text{ or } 23$$
$$\underline{t - 2k = 1 \text{ or } 3 \text{ or } 5 \text{ or } 15}$$

So $\qquad\qquad 4k = 344 \text{ or } 112 \text{ or } 64 \text{ or } 8$

$\qquad\qquad\quad\; k = 86 \text{ or } 28 \text{ or } 16 \text{ or } 2$

Making $\qquad\quad\; x = 94 \text{ or } 37 \text{ or } 26 \text{ or } 17$

$\qquad\qquad\quad\; y = 36 \text{ or } 13 \text{ or } 2 \text{ or } 3$

Now we can see that Mum could not be 94 if Bob were 36; and Bob cannot be 2 or 3.

Hence we are left with the one solution:

$$x = 37, y = 13$$

So Bob was 13 years old.

TYPICAL SOLUTION F

7. *The Funny Figures One Sees*

Let N be the required number.

$$N = 3a + 1 = 5b + 2 = 7c + 3 = 9d + 4$$

where a, b, c, and d are whole numbers.

Then $\qquad\qquad\qquad 7c - 9d = 1$

This is an indeterminate equation, but not of the simplest type. (See *Typical Solution D*.)

Divide through by 9:

$$c - \frac{2c}{9} - d = \frac{1}{9},$$ whence $\frac{2c+1}{9}$ is a whole number.

Now multiply the numerator by 4. This multiplier is chosen as being the lowest multiplier that will make the new coefficient of c one less or one more than a multiple of 9.

We get $\frac{8c+4}{9}$, which must also be a whole number.

This becomes $$c - \frac{c}{9} + \frac{4}{9}$$

Whence $\frac{c-4}{9}$ is a whole number, say t.

Then $\qquad\qquad\qquad c = 9t + 4$

Now $\qquad\qquad\qquad 5b + 2 = 7c + 3$

Substituting in this for c:
$$5b - 63t = 29$$

By similar process this gives $b = 63k + 31$, where k is a whole number.

Again $\qquad\qquad\qquad 3a + 1 = 5b + 2$

Substituting in this for b,
$$3a - 315k = 156$$

Simplifying this we get
$$a - 105k = 52$$

i.e. $\qquad\qquad\qquad a = 105k + 52$

Now $N = 3a + 1$, and N is in the 'one hundreds'.

So k must be zero, whence $a = 52$.

That gives us: $N = 157$.

TYPICAL SOLUTION G

8. *Trust a Postman to Know*

'Each dwelling faces its twin', so there must be an even number of houses, say $2n$ houses.

The distance between the first and last houses on each side will be $64(n-1)$ feet.

The distance between each house and its twin across the way is 112 feet.

The diagrams show the alternative routes:

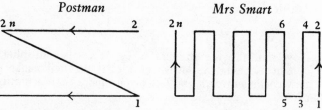

Postman   Mrs Smart

In a right-angled triangle, 'the square of the hypotenuse is equal to the sum of the squares of the other two sides'. Hence the total distance by the postman's new routine would be:

$$[128\,(n-1) + \sqrt{\{64^2\,(n-1)^2 + 112^2\}}] \quad \text{feet}$$

i.e. $16\{8n - 8 + \sqrt{(16n^2 - 32n + 65)}\}$ feet

And the total distance by Mrs Smart's route would be:

$$\{64\,(n-1) + 112n\} \quad \text{feet}$$

i.e. $16\,(11n - 4)$ feet

Equating the two distances, which are equal, and simplifying, we get: $n^2 - 8n + 7 = 0$
whence $n = 1 \text{ or } n = 7$

There are more than 2 houses, hence $n = 7$.
So there are 14 houses in Myrtle Drive.

TYPICAL SOLUTION H

10. No Seven Here

```
    2 x x
    3 x x
    -----
    5 x x
    x 4 x
    x x 3
    ---------
    x x x x x
```

The 1st product starts with 5, so the 1st multiplier must be 2. Inserting that figure, and since 3, the 3rd multiplier, gives a product ending in 3, we now have:

```
    2 A 1
    3 B 2
    -----
    5 x 2
    x 4 x
    x x 3
    ---------
    x x x x x
```

For ease of reference, letters A and B have been used.
As (2 A 1) multiplied by 2 gives (5 x 2), A is greater than 4.
Then, since the 2nd product has only 3 digits, B is less than 5.
Now consider the 2nd product. Within the limits which we have established for A and B, there are only two feasible alternatives:

$$B = 2, A = 7 \text{ or } B = 3, A = 8$$

But there can be no 'seven'; so B = 3, A = 8.
Hence we are left with:

$$281 \times 332 = 93292$$

TYPICAL SOLUTION I

11. Tim and His Brother Tom

Tom's age is x years, Tim's y years.
Working backwards from the statement of fact (i.e. 'Tom *was* two years younger etc.'), we get:
When Tim is twice his present age, Tim will be $2y$.
When Tom was 15 years younger than that, Tom was $(2y - 15)$.
When Tim was twice as old as that, Tim was $(4y - 30)$.
When Tom was as old as that, Tom was $(4y - 30)$.
Working backwards from the end, we get:
7 years ago, Tom was $(x - 7)$.
When Tim was 1 year older than that, Tim was $(x - 6)$.
When Tom was a third of that, Tom was $\dfrac{x - 6}{3}$.

When Tim was 3 years older than that, Tim was $\dfrac{x + 3}{3}$.

When Tom was 2 years younger than that, Tom was $\dfrac{x - 3}{3}$.

So we have $\qquad 4y - 30 = \dfrac{x - 3}{3}$

and also $\qquad x + y = 17$
Whence $\qquad x = 9, y = 8$.
So Tim is 8 years old.

TYPICAL SOLUTION J

13. No Catch in This

Let the three digits be x, y, and z in that order.

Then $$3z - 2x + 12y = 66$$

This is an indeterminate equation with three unknowns; in most cases this is handled on much the same lines as an indeterminate equation with two unknowns. In this case, however, there is a shorter method of solving it.

$$12y = 66 - (3z - 2x)$$

and $3z - 2x$ cannot be negative, hence y is less than 6. None of the digits can be 5, so y is less than 5.

$3z - 2x$ cannot exceed 25, i.e. $27 - 2$.

So $12y$ cannot be less than 41, whence y is greater than 3.

So $$y = 4$$

Thus $$3z - 2x = 18$$

This indeterminate equation can be solved at sight:

$$x = 0, z = 6 \text{ or } x = 3, z = 8$$

The wording of the first verse, coupled with the assurance that 'there's no catch in this', suggests that the first digit cannot be zero.

Hence the digits are 348.

TYPICAL SOLUTION K

14. Kim's Camera

The camera cost him exactly $\$y$.

He paid x dollar bills and x of each denomination of coins.

So altogether he paid:

$$x (100 + 25 + 10 + 5 + 1)\cent = 141x\cent$$

Hence $141x = 100y$ and y is less than 150.

So $$y = 141, x = 100$$

The camera cost him $\$141.00$.

TYPICAL SOLUTION L

15. The Old Soldier

He enlisted in 1890, so the date in question was not

before 1890.

Say the date was 1ABC, where A is 8 or 9. This date has actual value $1000 + 100A + 10B + C$.

The exchange of 2nd and 4th digits would give a new date, 1CBA, with actual value $1000 + 100C + 10B + A$.

Hence, as the new date is the later of the two,

$(1000 + 100C + 10B + A) -$

$$-(1000 + 100A + 10B + C) = 99$$

Whence $\qquad\qquad C - A = 1$

If $A = 9$, C would equal 10, which is impossible.

So $\qquad\qquad A = 8$ and $C = 9$

As the date was not before 1890, B must be 9.

Hence the date was 1899.

TYPICAL SOLUTION M

20. *How Many Marbles?*

They started with: Tam, $10x$; Tim, x; Tom, $2x$.

When Father came out, Tim had won a marbles, and either Tam or Tom had lost a.

The boy who grumbled could not have been Tim or Tom, so he must have been Tam. This is confirmed by the term 'third boy' as applying to Tim.

So it was Tam who had lost a marbles to Tim.

When Father came out, the position was:

$$\text{Tam: } 10x - a$$
$$\text{Tim: } \;\;x + a$$
$$\text{Tom: } 2x$$

So when the game ended, they had:

$$\text{Tam: } \;\;2x$$
$$\text{Tim: } 10x - 3$$
$$\text{Tom: } [13x - \{2x + (10x - 3)\}] = (x + 3)$$

But two of the boys 'ended up equal'.

So $\qquad 2x = (10x - 3)$, which is impossible

or $\qquad 10x - 3 = x + 3$, which is impossible

or $\qquad 2x = x + 3$, which is quite possible.

Hence, there being no other alternatives, $x = 3$.

So they had 39 marbles among them.

TYPICAL SOLUTION N

28. *The Beauties of Kalota*

Kassa said: Kissa, 23; Kessa, 22.

Kessa said: Kossa, 20; Kissa, 22.

Kissa said: Kassa, 22; Kossa, 19.

Taking the three together:

If Kissa was 23, then she was not 22.

So in that case Kossa was 20, and, as Kossa could not then be 19, Kassa was 22.

This would make Kessa's age 19.

If Kessa was 22, then Kissa was not 22 (i.e. there are no twins).

So in that case Kossa was 20, which again makes Kassa's age 22.

This is impossible, as there are no twins.

So Kissa was 23, and the four ages were:

Kissa: 23

Kossa: 20

Kassa: 22

Kessa: 19

TYPICAL SOLUTION N: PART II

28. *The Beauties of Kalota*

BY BOOLEAN ALGEBRA

Many problems of this type may be solved very neatly by the methods of Boolean Algebra, which is the basis of the mathematics of logic. Only a bare outline of the elements of this is possible here, but it may be of interest.

We adopt the convention that something 'true' has the value 1, and something 'false' the value 0. Using code symbols for the 'somethings', we can form expressions and equations that may be treated much the same as those in normal algebra.

A very simple example will show how the method is used. Say we have two statements about the name of a girl: one man said 'Betty Price,' the other 'Gwen Price,' and we are told that each made one mistake. Obviously Price was her surname, and her first name could not be either Betty or Gwen. Now let us see how this would be handled by Boolean Algebra.

We have only the two numerical values, 0 and 1. There's nothing more true than 'true': if, in the course of working, we derive any number greater than unity we must represent it as unity.

Let B stand for Betty, G for Gwen, and P for Price. Then we represent each statement in two ways:

Multiplication: If both B and P equal 1 (i.e., both true), then the product BP = 1. But if either B or P equals zero (i.e., false), then the product BP = 0.

Addition: If either B or P (or both) equals 1 (i.e., true), then B + P = 1.

So we can say, $(B + P)(G + P) = 1 \times 1 = 1$,
whence $BG + BP + GP + P^2 = 1$.

But BP = 0, GP = 0, and obviously BG = 0, so we are left with $P^2 = 1$, i.e., P = 1, which tells us that her surname was Price.

Now we can deal with the teaser of the Kalotan girls. Let our code be:

Kissa	— I	19 —	a
Kessa	— E	20 —	b
Kossa	— R	22 —	c
Kassa	— A	23 —	d

In code, then: Kassa said: Id, Ec
Kessa said: Rb, Ic
Kissa said: Ac, Ra

So we have:

Id + Ec = 1   and Id.Ec = 0
Rb + Ic = 1   and Rb.Ic = 0
Ac + Ra = 1   and Ac.Ra = 0

There are no twins, so any term of the form Ec.Ic **must** have 'value' zero.

No girl can have two ages, so terms of the form Id.Ic must have 'value' zero.

Now combine as $(Id + Ec)(Rb + Ic) = 1$

Whence

Id. Rb + Id. Ic + Ec. Rb + Ec. Ic = 1

Striking out terms with zero 'value', this becomes

Id. Rb + Ec. Rb = 1

Now bring in the third equation, as
$$(Id. Rb + Ec. Rb) (Ac + Ra) = 1$$
Whence
$$Ac. Id. Rb + Ac. Ec. Rb + Ra. Id. Rb + Ra. Ec. Rb = 1$$
Striking out terms with zero 'value', we are left with
$$Ac. Id. Rb = 1.$$
This shows that the ages were:

>Kassa: 22
>Kissa: 23
>Kossa: 20

So Kessa was 19.

TYPICAL SOLUTION O

24. *Those Licence Plates*

Represent their respective numbers, in conformity with the information, as:

>Paul: A B C D
>Dick: A D D E
>Hal : D B C A

Giving these their true values, from Hal's statement we get:
$$1000A + 100B + 10C + D + 1000A + 110D + E$$
$$= 1000 D + 100B + 10C + A$$
Whence $$1999A - 889D + E = 0$$
This is an 'indeterminate equation' with three unknowns, to be dealt with on somewhat similar lines to one with two unknowns.

Dividing by 889, we have:
$$2A + \frac{221A}{889} - D + \frac{E}{889} = 0$$
Whence
$$E = 889t - 221A \qquad D = 2A + t$$

Now if $A = 0$ and $t = 0$, then Dick's number would be 0000, and Paul and Hal would have the same numbers: this would have been a truly amazing coincidence and could not have escaped mention. So we may exclude that possibility.

So t must have minimum value 1 and must be positive;

A may well have value 0.

From $D = 2A + t$, since D must be less than 10, A cannot be greater than 4.

From $889t = 221A + E$, since A must be 1, 2, 3, or 4, and E can be any digit from 0 to 9, t cannot be greater than 1.

Hence $t = 1$

Thence we have:
$$E = 889 - 221A, \quad D = 2A + 1$$

E and D are both less than 10, so $A = 4$.

Whence $E = 5$ and $D = 9$

So Dick's number was 4995.

Answers

THE CAPITAL LETTER INDICATES THE 'TYPICAL SOLUTION' WHICH MAY PROVIDE A CLUE TO THE METHOD OF SOLVING THE PROBLEM

1. B. 2 miles per hour.
2. A. $3.12.
3. C. 30 years.
4. D. 4 males, 2 females.
5. E. 13 years.
6. A. 2 years older.
7. F. 157.
8. G. 14 houses.
9. A. 10 cigarettes.
10. H. 93292.
11. I. Tim is 8 years old.
12. A. 36 chickens.
13. J. 348.
14. K. $141.
15. L. 1899.
16. A. $7.00.
17. D. 8 cars, 4 horses, 3 bicycles.
18. A. She spent $23.00.
19. I. Jim was 35 years old.
20. M. 39 marbles.
21. N. Tuesday, June the 10th.
22.   1, 3, 9, and 27 kalens. The theoretical solution involves very advanced application of the Theory of Numbers.
23. A. 480 cars going north.
24. O. 4995.
25. D. Sal earned $47.47.
26. A. 23 cars had passed.

27. E. 5 oranges.
28. N. Kissa, 23; Kassa, 22; Kossa, 20; Kessa, 19.
29. A. Jim killed 4 birds.
30. A. Milly made 14 muffins.
31. B. 160 miles per hour.
32. A. The area was 84 square yards.
33. D. St Mark's, 27 seconds; St Mary's, 17 seconds.
34. A. $32.89.
35. I. Peg is his niece.
36. A. 1952.
37. D. At 6.36.
38. A. 20¢.
39. D. 13 @ 89¢, 19 @ $1.04.
40. N. Alf and Betty Cross, Charles and Amy Briggs, Bert and Clara Armstrong.
41. A. Jack had $1.60.
42. B. Roy had travelled 64 feet.
43. M. Ann was Len's wife.
44. J. Two 2¢ stamps.
45. A. The wrong entry was 45¢ for the second day.
46. H. 45353.
47. A. Mike's regular stride is 30 inches.
48. D. 80 geese.
49. A. 11 birds.
50.   The frontage was x yards. Then the depth was $\dfrac{120 - 4x}{3}$ yards. The depth being an exact number of yards, x must be a multiple of 3, say $x = 3k$. Then the area was $12k\,(10 - k)$ square yards, and k is obviously less than 10. Tabulating for whole-number values of k, from $k = 1$ to $k = 9$, we find that the maximum area is 300 square yards when $k = 5$. So the area of the lot was 300 square yards. (*Note*: This can be solved more readily by using elementary calculus.)

51. A. The kettle would have cost $4.25.
52. B. From where they started, Zirl was 102 miles.
53. A. He bought 6 ounces altogether.
54. D. The check was for $2.28.
55. A. 6 kids.
56. N. Elsie could not have been there.
57. J. 9317.
58. A. 72 stamps.
59.   5 becomes 80, 15 becomes 74, 20 becomes 71. The rule of transformation obviously operates in proportion, so $(20 + 10)$ becomes $(71 - 6)$, whence 30 becomes 65.
60. L. The correct change was $1.81.

61. If Jim bought 9, the outlay would be $3.80. If Joe bought 9, the outlay would be $3.60. So Jack bought 9, making the outlay $4.00.
62. G. The answer was 'Black and Red'.
63. B. 12 miles.
64. A. 1941.
65. N. Mr Penn was the lawyer, Mr Law the dentist, Mr Bankes the author, and Mr Tooth the banker.
66. J. The number was 4483.
67. D. A 5-dollar bill.
68. N. John Everard, Jim Lennox, Jack Best.
69. A. Les started with 19 cigarettes.
70. Any one of the boys bought x pint cartons and y quart cartons, whence $x + 2y = 5$. There are three alternative solutions to this:
$$x = 1 \text{ or } 3 \text{ or } 5$$
$$\text{with } y = 2 \text{ or } 1 \text{ or } 0$$
So each solution gives the purchase of one of the three boys. Tom bought the most pints, so he bought 5 pint cartons.
71. E. Tony was 10 years old.
72. C. The weight was 23 ounces.

73. A. The tie cost $2.50.
74. D. They had concealed 73 eggs.
75. B. Ben's speed was 27 miles per hour.
76. A. Gwen started with $26.00.
77. I. He died aged 12 years.
78. A. The height was 28 feet at the end of the 10th year.
79. G. He had to travel 5 feet.
80. D. 7 cats ate 'mousie meat'.
81. A. 209 voted for Kar.
82. L. 24¢ for bananas, 42¢ for pears.
83. H. $813 + 518 = 1331$.
84. N. King Kormon died in the year of Karan.
85. A. The weights were 9, 5, and 4 pounds.
86. D. 11 children and 38 dimes.
87. A. He started with $8.75.
88. I. The monkey weighed 5 pounds, whence we find that the rope was 20 feet long.
89. They must have been born in a leap year: Betty on January the 1st; Bella on February the 29th. The only leap year between 1933 and 1939 was 1936.
90. A. Sal made 52 collars.
91. I. Susan is 20 years old.
92. The minimum number 'smoking, with elbows on table' was $[46 - (57 - 37)] = 26$. The minimum number doing both, and also laughing, was $[43 - (57 - 26)]$

= 12. So the minimum number doing all three but *not* wearing glasses was [(57 — 8) — (57 — 12)] = 4. Hence, the required 'least number' was 4.

93. A. Bert dug 40 yards.

94. D. 101 letter-size; 45 legal-size.

95. A. 19¢ each.

96. A. 60 pupils altogether.

97. I. Jack is 11 years old.

98. A. Kurt's birthday was July the 16th.

99. E. Kurtis's age when his son was born was x years; Kalia had been with him y months. Then we get

$$5x - \frac{y\,(y+1)}{200} = 23.$$ Thence $x = 55$. Kurtis was 55 when his son was born.

100. A. Gwen's age is 19 years.

101. N. Larry was first, Steve second, and Ken third.

102.   Mike started with x¢, and won n games. If they played 7 games, $x = \dfrac{128 \times 555}{(128 - 3n)}$ and, as $128 - 3n$ is positive, n is less than 5. By trial we find $128 - 3n$ is not a factor of 128×555, so they did not play 7 games. As they played 6 games, $x = \dfrac{64 \times 555}{64 - 3n}$, in which case n is less than 4. By trial we find $64 - 3^3$ is a factor of 64×555, whence $x = 960$. So Mike started with $9.60 and ended up with $4.05.

103. A. He paid $22.95.

104.   There can be several alternative expressions for any number in this type of teaser. Examples are:

$$7 = 1 \times (4 + \sqrt{9}) \qquad = 1 \times (4 + 3)$$
$$15 = (1 + 9) \div \sqrt{.\overline{4}} \qquad = 10 \div \frac{2}{3}$$
$$28 = 4! + 1 + \sqrt{9} \qquad = (4 \times 3 \times 2 \times 1) + 1 + 3$$
$$30 = \Sigma\,(4) \times 1 \times \sqrt{9} \qquad = (4 + 3 + 2 + 1) \times 1 \times 3$$

105.   Bert lost $3 change, plus whatever the shoes had cost him.

106. A. 8 brothers and 4 sisters.

107. B. The distance was 3 miles.

108. D. 6 saucers.

109. A. The purse contained $1.40.

110. I. John is 8 years old.

111.   The question could be: 'If, in my next question, I were to ask you, "Does the left road lead to Knokado?" would you answer "Yes"?'. Then, if in fact the left road does lead to Knokado, she is bound to say 'No'; and if it

does not lead to Knokado, she is bound to say 'Yes'. Remember, she could not lie in reply to two consecutive questions.

112. D. The taxi fare was $1.95.

113. E. 5 persons had subscribed.

114. N. Joe Leary, with 3 sons.

115.   The ages of brothers are x and $(x + 10)$ years, and Sue's age is y years. Then $y + 5x = 42$ or $y + 5x = 52$, x and y being whole numbers as it was their birthday.
Also, $y = x + 8$ or $y = x + 18$
so $x = y - 8$ or $x = y - 18$.
Substituting these values for x in the two original alternative equations, we get $6y = 82$ or 92 or 132 or 142. A whole-number value for y is obtained only from $6y = 132$ when $y = 22$. So Sue was 22 years old.

116.   Amelia is x, but feels y years old. From her final remark, Sally looks $(x - 20)$ years old. If Amelia had been $(x - 20)$ ten years ago, she would now be $(x - 10)$ years old. If Sally were as old as Amelia feels, Sally would be y years old; but that would be $(x - 10)$ $+ 20$ years old. So $y = x + 10$. Hence Amelia feels 10 years older than she is.

117. A. Aunt Emily was 42 years old.

118. L.   The number has digits x, y, a, and b in that order. So
& O. $1999x + 190y - 80a - 998b = 39$. Dividing by 10, we get: $x = 10t + 2b + 1$, t being a whole number. Now x and b are both less than 10, so if
$$b < 5, x = 2b + 1$$
and if $b > 4, x = 2b - 9$
Say $b < 5$, and substitute for x in the equation. Then we get $300b + 19y - 8a = -196$.
Here b must be   0,   1,   2,   3,   or   4
with $8a - 19y =$ 196, 496, 796, 1096, or 1396.
But $8a - 19y$ cannot exceed 72, so none of these is acceptable.
Say $b > 4$. Then $300b + 19y - 8a = 1803$.
Then b must be   5,   6,   7,   8,   or   9,
with $19y - 8a =$ 303, 3, -297, -597, or -897
But $19y - 8a$ cannot exceed 171, so $b = 6$, and $19y - 8a = 3$. Then from $19y - 8a = 3$, we have $y = 1$, $a = 2$. Also $x = 2b - 9$, so $x = 3$. Hence the number was 3126.

119. E. Betty was 10 years old.

120. D. He paid $1.33.

121. A. She started with and spent $6.20.

122.   The speaker was 63 years old.

123. D. Aunt Amelia was 43 years old.

124. A. $9.

125. B. He had walked for 26 minutes.
126.   Weigh any 3 against any 3. *If they balance*, the 'fake'
        is the lighter of the other 3. Weigh one of these against
        one; if they balance, the 'fake' is the odd coin; if they
        do not balance, the 'fake' is the lighter of the two being
        weighed. *If they do not balance*, the 'fake' is the
        lighter coin among the lighter 3. Proceed with these
        lighter 3 as above.
127. A. The wall was 10 feet long.
128. N. Tam, Monday; Tim, Tuesday; Tom, Friday.
129. E. The car number is 2205.
130. A. 27,000 marched the previous day.
131. G.

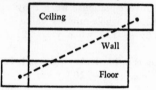

The distance was 40 feet.

132. I. Her brother was 14 years old.
133.   He bought 7 cigars.
134. G. 26 feet.
135. A. Betty's stride was 16 inches.
136. D. He spent $6 on Friday.
137. A. 21 skirts in a day.
138. H.

```
    1 2 3
    1 6 3
    -----
    3 6 9
  7 3 8
1 2 3
---------
2 0 0 4 9
```

139. A. The population of Neepawa was 2292.
140. A. Bert was 44 years old.
141. C. Uncle Jim had $1.80 in his hand.
142. A. 22 birds.
143. E. 20 bananas altogether.
144. A. Mike earned $483 that month.
145. A. $2.75 and $3.25.
146. L. She was 73 years old.
147.   He ended up with x red, $(x - 2)$ green, $(13 - 2x)$
        blue. $(13 - 2x)$ cannot be 2; and x cannot be 2, as he
        did not have more than 7 of one colour. So $x - 2 = 2$.
        Hence he ended up with 4 red, 2 green, and 5 blue.
        So he started with 5 red and 7 blue, or with 7 red
        and 5 blue: either way, he started with 3 green.
148. B. Mike caught Steve in 1½ hours.

149. E. His mother was 41, and Pam was 8 years old.
150. A. 15 cookies.